Origin, Evolution and Biogeographic History of South American Turtles

Springer Earth System Sciences

For further volumes:
http://www.springer.com/series/10178

Marcelo S. de la Fuente • Juliana Sterli
Ignacio Maniel

Origin, Evolution and Biogeographic History of South American Turtles

Springer

Marcelo S. de la Fuente
CONICET
Museo de Historia Natural de San Rafael
Parque Mariano Moreno S/N
San Rafael, Mendoza, Argentina

Juliana Sterli
CONICET
Museo Paleontológico Egidio Feruglio
Trelew, Chubut, Argentina

Ignacio Maniel
CONICET
Fundación de Historia Natural
Felix de Azara
Universidad Maimonides
Buenos Aires, Argentina

ISBN 978-3-319-37640-0 ISBN 978-3-319-00518-8 (eBook)
DOI 10.1007/978-3-319-00518-8
Springer Cham Heidelberg New York Dordrecht London

Printed on acid-free paper

Springer is part of Springer Science+Business Media (www.springer.com)

This monograph is dedicated to the memory of the Argentinean paleontologist Prof. Dr. Rosendo Pascual (1925–2012). Rosendo Pascual, in addition to serving the science generously, has influenced with his ideas three generations of vertebrate paleontologistsin Argentina for the last 50 years.

Preface

This volume is partly based on a lecture given in 2011 on the Symposium "Seaways and landbridges: Southern Hemisphere biogeographic connections through time" held in Argentina, and it is an attempt to show the knowledge of South American turtles, as the result of many years of work on turtle systematic as vertebrate paleontologists. The aim of this monograph is to present a concise summary on the origin, evolution, and biogeographic history of South American continental turtles and tortoises. We hope this monograph will be interesting for students of general Paleontology and Zoology, as well as for herpetologists, and also for general readers interested in South American turtles.

As far as possible we have tried to make a balance in the treatment of the extant and the less known extinct South American turtle species. Each chapter deals with different groups of turtles and tortoises, and they are ordered accordingly with the timing in the differentiation and integration of each clade in South America. All the South American species mentioned in the text are listed with systematic comments and with either their distribution (for extant species) or their geographic and stratigraphic procedence data (for extinct forms).

This monograph was possible thanks to many people and funding. In this regard, we would like to thank S. Casadio, L. Salgado, Rodolfo Coria (Universidad Nacional de Río Negro, General Roca, Argentina), and Miguel Griffin (Universidad Nacional de La Plata, La Plata, Argentina) for the organization of the Symposium "Seaways and landbridges: Southern Hemisphere biogeographic connections through time" held in General Roca, Río Negro Province, Argentina (April 26 –May 1, 2011), and J. Rabassa (CADIC, Ushuaia, Tierra del Fuego, Argentina) for arranging the publication of many lectures and discussions resulting from this symposium, some of them in the SpringerBriefs in Earth Sciences Series. We would also like to recognize the assistance of Raven Garvin (University of California at Davis) and Jorge Rabassa for English grammar revision and helpful comments on an early draft. The 16 artwork illustrations were made by Jorge González. The other sketch maps and diagrams are our own. This project was partially founded by the PIP-CONICET Grant 0795 to Marcelo S. de la Fuente.

San Rafael, Argentina — Marcelo S. de la Fuente
Trelew, Argentina — Juliana Sterli
Buenos Aires, Argentina — Ignacio Maniel

Contents

Chapter 1
Introduction

Abstract The continental chelonian fauna of South America is integrated by almost 46 extant species assigned to seven different clades (Emydidae, Geoemydidae, Chelydridae, Kinosternidae, Testudinidae, Chelidae, and Podocnemididae) and 48 extinct species referred to ten different clades (Emydidae, Trionychidae, Testudinidae, Chelidae, Podocnemididae, Podocnemidinura, Bothremydidae, Meiolaniformes, Platychelyidae, and Australochelyidae). The chelonian fauna of South America has been exposed to multiple and different processes since the Late Triassic (e.g., origin, dispersion, vicariance, and extinction) that shaped their modern distribution in the continent. The fossil record of turtles shows a different timing in the integration of each clade to South America.

Keywords Testudinata • Mesozoic • Cenozoic • South America • Evolution • Systematic • Biogeography

The continental chelonian fauna of South America is integrated by almost 46 extant species assigned to seven different families (Emydidae, Geoemydidae, Chelydridae, Kinosternidae, Testudinidae, Chelidae, and Podocnemididae). Even though the extant representatives live in several South American continental basins (i.e., Amazonian, Orinoco, Paraguay, San Francisco), the clades have undergone through different biogeographic histories produced by vicariant and dispersal events. The fossil record of South American turtles, consequently, shows a different timing in the integration of each turtle clade to South America.

Regarding the specific diversity of the South American chelonian fauna, the oldest clades with extant representatives (Chelidae and Podocnemididae) show the highest specific richness in comparison with other clades (e.g., Emydidae, Chelydridae, Kinosternidae, Testudinidae). Particularly, Chelidae is represented by 36 species (22 extant and 14 extinct) referred to 14 genera (*Acanthochelys*, *Prochelidella*, *Platemys*, *Mesoclemmys*, *Bonapartemys*, *Linderochelys*, *Palaeophrynops*, *Phrynops*, *Rhinemys*, *Salamanchelys*, *Chelus*, *Lomalatochelys*, *Hydromedusa*, and

M.S. de la Fuente et al., *Origin, Evolution and Biogeographic History of South American Turtles*, Springer Earth System Sciences, DOI 10.1007/978-3-319-00518-8_1,

Yaminuechelys). Likewise, Podocnemididae includes 25 species (7 extant and 18 extinct), assigned to 13 genera (*Bauruemys*, *Carbonemys*, *Camberemys*, *Cerrejonemys*, *Peiropemys*, *Lapparentemys*, *Pricemys*, *Podocnemis*, *Caninemys*, *Peltocephalus*, *Bairdemys*, *Stupendemys*, and *Roxochelys*). Contrarily to this relatively high specific and generic diversity, the aquatic cryptodiran turtles (Emydidae, Geoemydidae, Chelydridae, Kinosternidae) and tortoises (Testudinidae) are represented by only one genus per clade (*Trachemys*, *Rhinoclemmys Chelydra*, *Kinosternon*, and *Chelonoidis*, respectively). The specific richness is also smaller in aquatic cryptodiran (Emydidae = 4 species, Geoemydidae = 5 species, Chelydridae = 1 species, Kinosternidae = 3 species) and also in true tortoises (Testudinidae), which include 8 species (4 extant and 4 extinct) (Fig. 1.1).

This strong difference in specific and generic diversity between the aquatic cryptodiran (plus the tortoises) and the eupleurodiran could be correlated to the timing of their origin or arrival to South America. Although the chelids and the ancestors of podocnemidid (basal pelomedusoids) had been originated at least as early as the Early Cretaceous in southern and northern Gondwana, respectively (Broin 1987, 1988; Broin and de la Fuente 1993), the testudinids and the aquatic cryptodires (emydids, geoemydids, chelydrids, kinosternids, trionychids) arrived to South America from other continents by dispersion. Africa or Central America has been proposed as the continental source for the Testudinidae (Simpson 1942, 1943; Auffenberg 1971; de la Fuente 1988, 1997; Gerlach 2001; Le et al. 2006) and Central America for the aquatic cryptodiran turtles (Vanzolini and Heyer 1985; Seidel and Jackson 1990; Vanzolini 1995; de la Fuente et al. 2002; Le and McCord 2008; Fritz et al. 2012).

On the contrary to some chelonian clades with extant representatives, fully extinct clades have more ancient origins. Whereas meiolaniforms had differentiated at least in the Early Cretaceous in the southern area of Gondwana (Sterli and de la Fuente 2012) and probably colonized Laurasia by dispersal, the Jurassic and Triassic chelonians had a widespread Pangaean distribution. Meiolaniforms are represented in South America by 4 extinct species (*Chubutemys capelloi* Gaffney et al. 2007, *Patagoniaemys gasparinae* Sterli and de la Fuente 2011, *Peligrochelys walshae* Sterli and de la Fuente 2012, and *Niolamia argentina* Ameghino 1899). Only two panpleurodiran species of Tithonian and Valanginian ages are known (*Notoemys laticentralis* Cattoi and Freiberg 1961 and *Notoemys zapatocaensis* Cadena and Gaffney 2005). Finally, the oldest South American representatives are the Middle Jurassic *Condorchelys antiqua* Sterli 2008 and the Upper Triassic *Palaeochersis talampayensis* Rougier, de la Fuente and Arcucci 1995.

Fig. 1.1 (continued) *Chubutemys*. *Co*, *Condorchelys*. CRET, Cretaceous. *Eu*. *Euraxemys*. *H*. *Hydromedusa*. JUR, Jurassic. *L*, *Lomalatachelys*. *La*, *Lapparentemys*. *Li*, *Linderochelys*. *N*, *Notoemys*. NEO, Neogene. *Ni*, *Niolamia*. *P*, *Prochelidella*. *Pa*, *Palaeophrynops*. *Pat*, *Patagoniaemys*. PALEO, *Paleogene*. *Pe*, *Peltocephalus*. *Pei*, *Peiropemys*. *Pel*, *Peligrochelys*. *Pl*, *Platemys*. *Po*, *Podocnemis*. *Por*, *Portezueloemys*. *Pri*, *Pricemys*. *R*, *Roxochelys*. *Rh*, *Rhinemys*. *S*, *Salamanchelys*. *St*, *Stupendemys*. TR, Triassic. *Y*, *Yaminuechelys*. Summary table of taxa analyzed. Each taxon is associated to the clade to which it was assigned and the period in which it was found. Numbers indicate million years ago

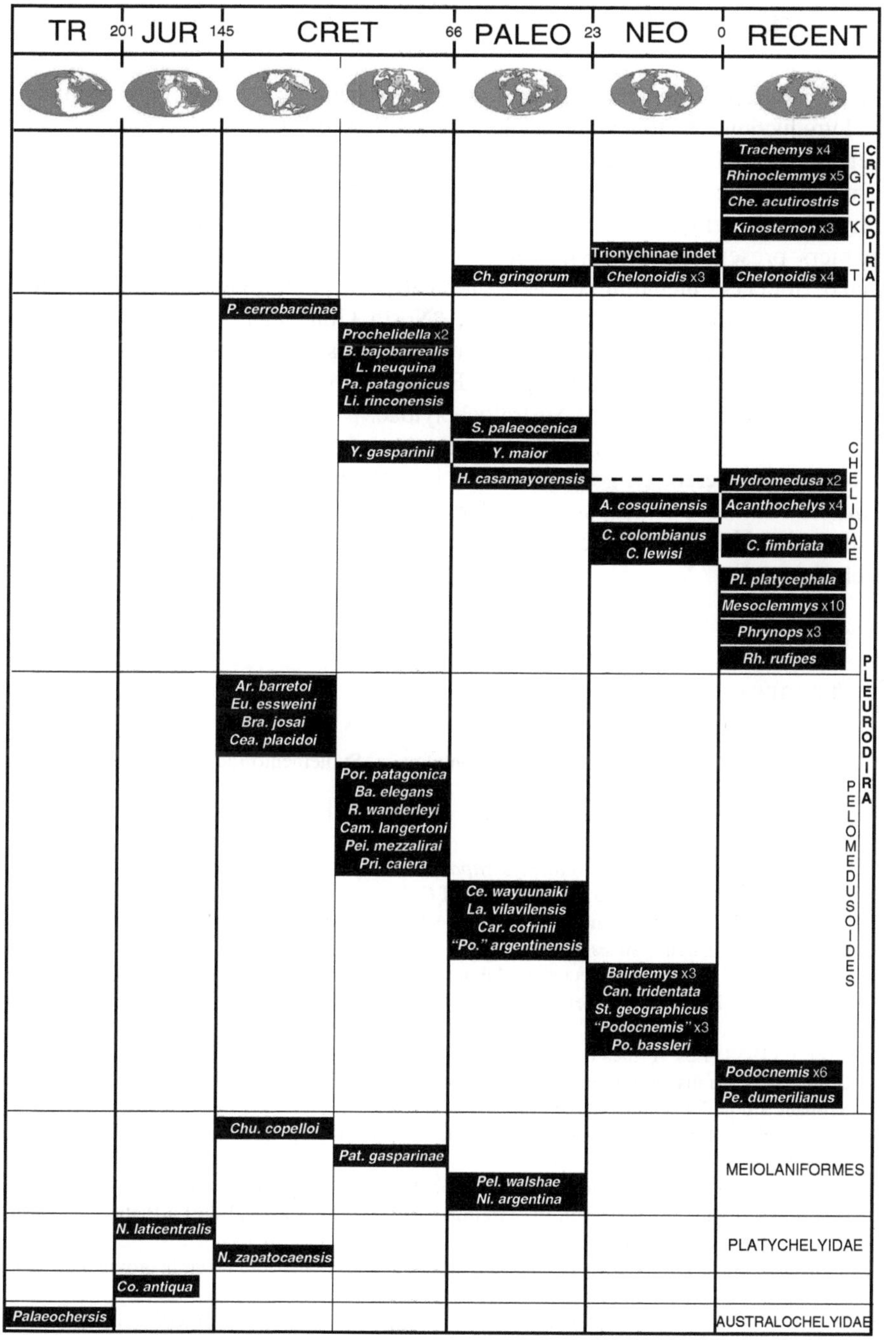

Fig. 1.1 Stratigraphic distribution of taxa analyzed in the text. *A*, *Acanthochelys*. *Ar*, *Araripemys*. *B*, *Bonapartemys*. *Ba*, *Bauremys*. *Bra*, *Brasilemys*. *C*, *Chelus*. *Can*, *Caninemys*. *Cam*, *Cambaremys*. *Car*, *Carbonemys*. *Ce*, *Cerrejonemys*. *Cea*, *Cearachelys*. *Ch*, *Chelonoidis*. *Che*, *Chelydra*. *Chu*,

1.1 Organization

This monograph is divided in different sections (chapters). Chapter 1 corresponds to the Introduction. The following chapters are arranged by clades and ordered following their chronological arrival time or differentiation in South America. Chapter 2 is named "Aquatic Cryptodiran Turtles: The Most Recent Island Hoppers" including Emydidae, Geoemydidae, Chelydridae, Kinosternidae, and Trionychidae. The other chapters present the following development: Chap. 3, "Tortoises: The Oldest Island Hoppers" including Testudinidae; Chap. 4, "South Gondwana Pleurodiran Turtles," including Chelidae; Chap. 5, "North Gondwana Pleurodiran Turtles," including Pelomedusoides; Chap. 6, "Meiolaniforms: An Extinct Lineage of Turtles of Gondwanan Origin"; and Chap. 7, "Early Differentiation of Mesozoic Turtles," including Platychelyidae and Australochelyidae.

Each section includes a brief abstract, a general characterization of the clade or clades considered, the genera included in each clade, and a summary of all South American extant or extinct species referred to a particular clade of turtle. Finally, a discussion about systematic and biogeographic topics is provided at the end of each chapter.

References

Ameghino F (1899) Sinopsis geológica paleontológica. Suplemento (adiciones y correcciones). Censo Nacional:1–13. La Plata.

Auffenberg W (1971) A new fossil tortoise, with remarks on the origin of South American Testudinines. Copeia 1:106–117

Cadena E, Gaffney ES (2005) *Notoemys zapatocaensis*, a new side-necked turtle (Pleurodura: Platychelyidae) from the Early Cretaceous of Colombia. Am Mus Novit 3470:1–19

Cattoi N, Freiberg M (1961) Nuevo hallazgo de Chelonia extinguidos en la República Argentina. Physis 22:202, Buenos Aires

de Broin F (1987) The Late Cretaceous Fauna of Los Alamitos, Patagonia, Argentina. Part IV, Chelonia. Revista Museo Argentino Ciencias Naturales 'Bernardino Rivadavia'. Paleontología 3:131–139

de Broin F (1988) Les Tortues et le Gondwana. Examen des rapport entre le fractionemment du Gondwana et la dispersión géographique des tortues pleurodires àpartir de Crétacé. Stud Geol Salmant Studia Palaeochel 2:103–142

de Broin F, de la Fuente MS (1993) Les tortues fossiles d'Argentine: synthèse. Ann Paléontol 79:169–232

de la Fuente MS (1988) La tortugas Chelidae (Pleurodira) y Testudinidae (Cryptodira) del Cenozoico argentinos. Doctoral dissertation, Universidad Nacional de La Plata, Facultad de Ciencias Naturales y Museo, La Plata

de la Fuente MS (1997) Las tortugas pleistocenas del extremo meridional de la provinciade Santa Fé, Argentina. Stud Geol Salmant 33:67–90

de la Fuente MS, Noriega JI, Piña CI (2002) *Trachemys dorbigni* (Duméril y Bibron, 1835) (Cryptodira: Emydidae) en el Pleistoceno tardío de la provincia de Entre Ríos, Argentina. Cuad Herpetol 16:65–72

Fritz U, Stuckas H, Vargas-Ramírez M, Hundsdörfer AK, Maran J, Päckert M (2012) Molecular phylogeny of Central and South American slider turtles: implications for biogeography and systematics (Testudines: Emydidae: *Trachemys*). J Zool Syst Evol Res 50:125–136

Gaffney ES, Rich TH, Vickers-Rich P, Constantine A, Vacca R, Kool L (2007) *Chubutemys*, a new eucryptodiran turtle from the Early Cretaceous of Argentina, and the relationships of the Meiolaniidae. Am Mus Nov 3599:1–35

Gerlach J (2001) Tortoise phylogeny and the '*Geochelone*' problem. Phelsuma 9:1–24

Le M, McCord W (2008) Phylogenetic relationships and biogeographical history of the genus *Rhinoclemmys* Fitzinger, 1835 and the monophyly of the turtle family Geoemydidae (Testudines: Testudinoidea). Zool J Linn Soc 153:751–767

Le M, Raxworthy CJ, McCord WP, Mertz L (2006) A molecular phylogeny of tortoises (Testudines: Testudinidae) based on mitochondrial and nuclear genes. Mol Phylogenet Evol 40:517–531

Rougier G, de la Fuente MS, Arcucci A (1995) Late Triassic turtles from South America. Science 268:855–858

Seidel ME, Jackson DR (1990) Evolution and fossil relationship of slider turtles. In: Gibbons JW (ed) Life history and ecology of the slider turtle. Smithsonian Institution Press, Washington, DC

Simpson GG (1942) A Miocene tortoise from Patagonia. Am Mus Novit 1209:1–6

Simpson GG (1943) Turtles and the origin of the fauna of Latin America. Am J Sci 241:413–429

Sterli J (2008) A new, nearly complete stem turtle from the Jurassic of South America with implications for turtle evolution. Biol Lett 4:286–289

Sterli J, de la Fuente MS (2011) A new turtle from the La Colonia Formation (Campanian–Maastrichtian), Patagonia, Argentina, with remarks on the evolution of the vertebral column in turtles. Palaeontology 54:63–78

Sterli J, de la Fuente MS (2012) New evidence from the Palaeocene of Patagonia (Argentina) on the evolution and palaeobiogeography of Meiolaniformes (Testudinata, new taxon name). J Syst Palaeont. doi:10.1080/14772019.2012.708674

Vanzolini PE (1995) A new species of turtle genus *Trachemys* from the State of Maranhao, Brazil (Testudines, Emydidae). Rev Brasil Biol 55:111–125

Vanzolini PE, Heyer WR (1985) The American Herpetofauna and the interchange. In: Webb SD, Stheli FG (eds) The Great American biotic interchange. Plenum Press, New York/London

References

Chapter 2
Aquatic Cryptodiran Turtles: The Most Recent Island Hoppers

Abstract Aquatic cryptodiran turtles of the clades Emydidae, Chelydridae, Geoemydidae, Kinosternidae, and Trionychidae are the most recent Mesoamerican immigrants to South America, arriving in various waves from the Miocene 12 million years ago (prior to the emergence of the Panamanian Isthmus) to the Plio-Pleistocene and Holocene (after the establishment of this isthmus). With the exception of Trionychidae, the remaining cryptodiran clades are still present in the South American chelonian fauna.

Keywords Mesoamerican immigrants • Miocene • Pliocene • Pleistocene • Panamanian isthmus

2.1 Emydidae Rafinesque 1815

Emydidae, also known as the New World pond turtles, includes 10 genera and around 51 species (see Rhodin et al. 2010) and is restricted to the Western Hemisphere with the exception of two species (Fig. 2.1a). This clade is closely related to Testudinidae (true tortoises), and they were included together with Geoemydidae in the clade Testudinoidea. Emydidae thought to have originated in North America where nine of the ten genera currently occur. This clade is clearly monophyletic (Bickham et al. 1996) and has been divided in two clades by Gaffney and Meylan (1988): Deirochelyidae, which includes predominantly aquatic species, and Emydinae, which includes both aquatic and terrestrial species. The major clades have been identified by Bickham et al. (1996) phylogenetic analysis. According to Ernst and Barbour (1989), Emydidae are characterized by the presence of a contact between the angular bone and Meckel's cartilage, a narrow basioccipital that is separated from the paracapsular sac and pterygoid that does not form the floor of the scala tympani, by the presence of a double articulation between the centra of the 5th and 6th cervical vertebrae, and placement of the seams between the 12th marginal scute

M.S. de la Fuente et al., *Origin, Evolution and Biogeographic History of South American Turtles*, Springer Earth System Sciences, DOI 10.1007/978-3-319-00518-8_2,

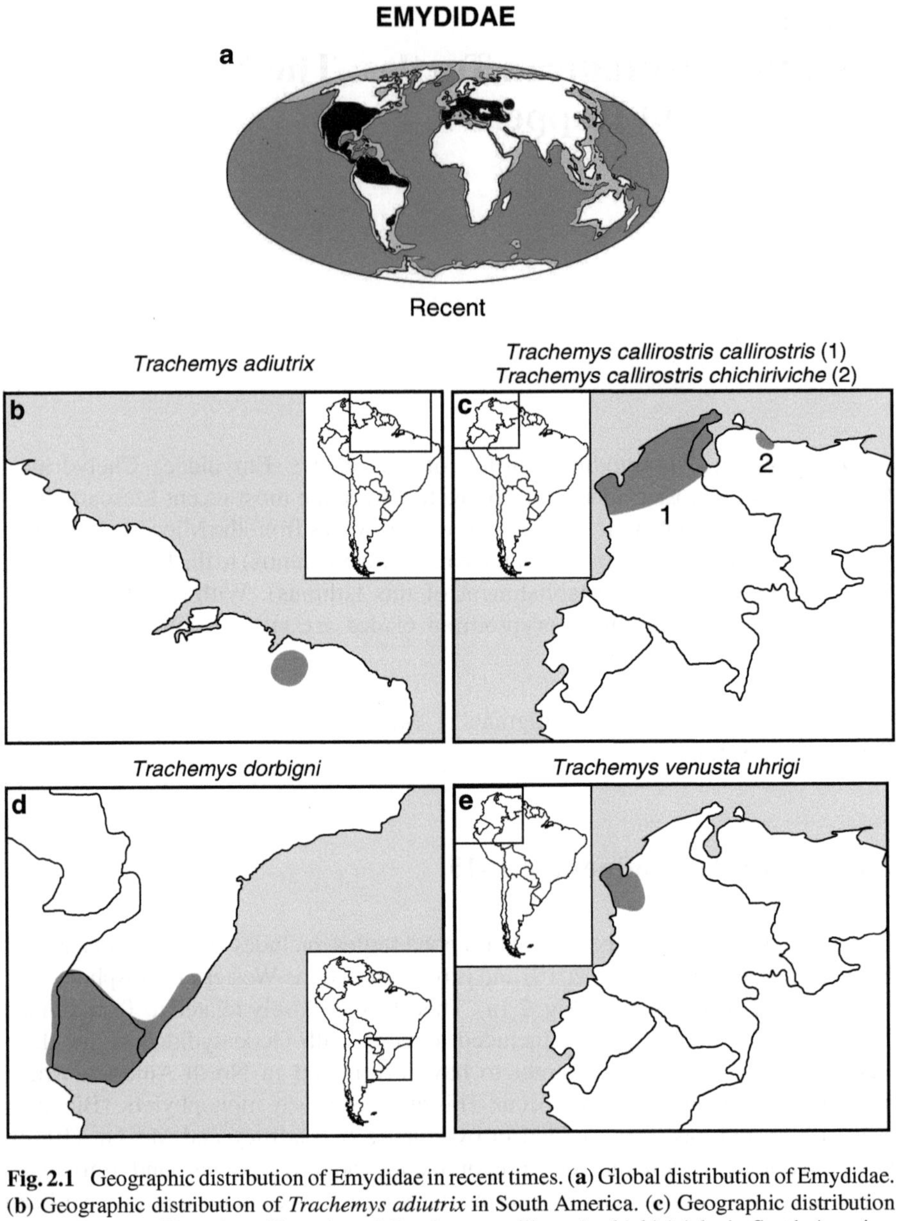

Fig. 2.1 Geographic distribution of Emydidae in recent times. (**a**) Global distribution of Emydidae. (**b**) Geographic distribution of *Trachemys adiutrix* in South America. (**c**) Geographic distribution of *Trachemys callirostris callirostris* and *Trachemys callirostris chichiriviche* in South America. (**d**) Geographic distribution of *Trachemys dorbigni* in South America. (**e**) Geographic distribution of *Trachemys venusta uhrigi* in South America

and the last vertebral over the pygal bone. A detailed diagnosis of Emydidae based on skull and lower jaw anatomy has been provided by Gaffney (1975).

Genera included: *Actinemys* Agassiz 1857, *Chrysemys* Gray 1844, *Clemmys* Ritgen 1828, *Deirochelys* Agassiz 1857, *Emydoidea* Gray 1870, *Emys* Duméril

1806, *Glyptemys* Agassiz 1857, *Graptemys* Agassiz 1857, *Malaclemys* Gray 1844, *Pseudemys* Gray 1856a, *Pseudograptemys* Loomis 1904, *Psilosemys* Hutchison 2013, *Terrapene* Merrem 1820, and *Trachemys* Agassiz 1857.

South American Taxa

2.1.1 *Trachemys* Agassiz 1857

Type species: *Emys troosti* Holbrook 1836.

Species included: *T. adiutrix* Vanzolini 1995, *T. angusta* (Barbour and Carr 1940), *T. callirostris* (Gray 1856b), *T. decorata* (Barbour and Carr 1940), *T. decussata* Griffith and Pidgeon 1830, *T. dorbigni* (Duméril and Bibron 1835), *T. emolli* (Legler 1990), *T. taylori* (Legler 1960), *T. gaigeae* (Hartweg 1939), *T. ornata* (Gray 1830 in Griffith and Pidgeon 1830), *T. scripta* (Schoepff 1792), *T. stejnegeri* Schmidt 1928, *T. terrapin* (Lacépède 1788), *T. venusta* (Gray 1856b), and *T. yaquia* (Legler and Webb 1970).

Trachemys adiutrix Vanzolini 1995

Holotype: Museu de Zoologia Universidade de Sao Paulo, Sao Paulo, Brazil. MZUSP 3224.

Diagnosis: See Vanzolini (1995).

Distribution: This species occurs in the vicinity of dune fields of the Lençois Maranhenses, Santo Amaro locality, Piaui, Maranhao, Brazil (Vanzolini 1995) (Fig. 2.1b).

Comments: This species was named by Vanzolini (1995) based on the holotype and five other specimens. In order to validate the specific entity of *T. adiutrix*, Vanzolini has made extensive comparisons among all South American species belonging to genus *Trachemys*. *Trachemys adiutrix* differs from *T. dorbigni* in overall shape, the shape of the first vertebral scale, plastral scale proportions measured in midline, and color patterns. According to Vanzolini (1995), *T. adiutrix* differs from northern South American subspecies of *T. callirostris* (Gray 1856b), *T. c. callirostris* and *T. c. chichiriviche* (Pritchard and Trebbau 1984), in coloration pattern, a more flared carapace, a pointed behind, and in the shape of the second vertebral scale. Recently, Fritz et al. (2012) referred *adiutrix* as a subspecies of *Trachemys dorbigni* based on molecular analyses.

Trachemys callirostris (Gray 1856b) (Fig. 2.2)

Holotype: British Museum of Natural History, London, United Kingdom, BMNH 1947.3.4.87.

Diagnosis: See Pritchard and Trebbau (1984).

Distribution: *Trachemys callirostris callirostris* is distributed across northern Colombia in the middle and lower Magdalena River and its main tributaries, the drainage of lower Sinú River (Ernst 2003), and drainage in the western region of

Fig. 2.2 Drawing of *Trachemys callirostris callirostris*

Lake Maracaibo, Zulia Province, Venezuela (Pritchard and Trebbau 1984). *Trachemys callirostris chichiriviche* occurs along river margins, restricted to the area along the Caribbean coast of northern Venezuela (Pritchard and Trebbau 1984) (Fig. 2.1c).

Comments: This species was originally described as *Emys callirostris* by Gray in 1856b and referred to the genus *Pseudemys* by Cope (1887), to *Callichelys* by Gray (1863a), and described as a subspecies of *Chrysemys ornata* by Boulenger (1989). More recently, Pritchard and Trebbau (1984) recognized *callirostris* as a subspecies of *Pseudemys scripta* and named and described *Pseudemys scripta chichiriviche*. Finally, Seidel (2002) designated *callirostris* as a species of *Trachemys* based on morphological data and named the subspecies *Trachemys callirostris callirostris* and *Trachemys callirostris chichiriviche*. This taxonomic decision is retained in Fritz and Havas (2007) and Rhodin et al. (2010). Detailed information about the ecology and biology of both subspecies has been published by Pritchard and Trebbau (1984) and Bock et al. (2010). Recently, Fritz et al. (2012) referred to this species as a subspecies of *T. ornata*.

Trachemys dorbigni Duméril and Bibron 1835

Holotype: Muséum National d´Histoire Naturelle de Paris, France. MNHN 9221.

Definition: See Ernst and Barbour (1989) and Seidel (1989).

Distribution: This species ranges from São Luis, Maranhao, and the Rio Guaiba drainage near Pôrto Alegre in Rio Grande do Sul, Brazil, to Paraguay, Uruguay, and northeastern Argentina through the Paraguay and Uruguay river drainages (Ernst and Barbour 1989) (Fig. 2.1d).

Comments: Although several authors (e.g., Moll and Legler 1971; Pritchard 1979; Iverson 1985; Alderton 1988) included *T. dorbigni* as a subspecies of *T. scripta*, others including Seidel (1990), Cabrera (1995), and Vanzolini (1995) noted that although this species is closely related to the *T. scripta* complex, its disjointed range relative to that of other extant *Trachemys* species allows its treatment as separate species. Likewise, Seidel (1990), Cabrera (1995), and Vanzolini (1995) argued that this species isolation precludes gene flow between northern and southern populations. In addition, Freiberg (1969) based on carapace coloration and morphological characters including plastral lobe proportions and epidermal scutes recognized two subspecies of *Trachemys dorbigni*: *T. dorbigni dorbigni* (restricted to Uruguay and northeastern Argentina) and *T. dorbigni brasiliensis* (restricted to southeastern Brazil). However, the characters that permit the distinction between subspecies are not evident. This is in agreement with the research made by del Barco and Larriera (1991), who did not recognize these subspecies and accepted *T. dorbigni* as a valid species. The fossil record of *Trachemys dorbigni* is limited to the Late Pleistocene (the Arroyo Feliciano Formation, in Entre Ríos Province; de la Fuente et al. 2002) and to the Late Pleistocene–Early Holocene (the Tezanos Pinto Formation, in the province of Córdoba Province, Argentina (Cabrera and Luna 2011)).

Trachemys venusta Gray 1856b

Syntypes: British Museum of Natural History, London, United Kingdom. BMNH 1839. 12.26.70, BMNH 1844.12.7.3, BMNH 1845.8.5.26, BMNH 10.20.19, 1848.7.28.25, 1849.12.7.4; the specimen BMNH 1845 was designed as the lectotype by Smith and Smith (1979).

Diagnosis: See original description in Gray 1856b and see also McCord et al. (2010).

Distribution: Northwestern Veracruz to the Yucatán Peninsula, Mexico, along the Atlantic Ocean coastal plain to northwestern Colombia (Iverson 1992).

Trachemys venusta uhrigi McCord, Joseph-Ouni, Hagen and Blanck 2010

Distribution: Northwestern Caribbean, coastal Honduras eastward to the lower Patuca River drainage, in the region of northeastern Caribbean coastal Honduras, Costa Rica, Panama, and into the lower drainages of the Atrato River of northwestern Colombia (McCord et al. 2010) (Fig. 2.1e).

Diagnosis: See McCord et al. (2010).

Comments: The Mesoamerican slider has had a complex taxonomic history since it was first described as *Emys venusta* by Gray in 1856b (see Fritz and Havas 2007). During the last century (from 1971 to 1992), it has been referred to as a subspecies of *Trachemys scripta* (as *Pseudemys* in Moll and Legler 1971, as *Chrysemys* in Henderson and Hoevers 1975, as *Trachemys* in Iverson 1985, 1992). Recently, Seidel (2002) elevated *venusta* to species rank as *Trachemys venusta* (Gray 1856b). Only one subspecies of *venusta* (*Trachemys venusta uhrigi*) is found in South America, distributed in the Antioquía and Choco departments of Colombia (McCord et al. 2010). More recently, Fritz et al. (2012) designated *venusta* as a subspecies of *T. ornata*.

2.2 Geoemydidae Theobald 1868

Geoemydidae is the largest turtle clade in the world (73 species following Spinks et al. 2004), representing 25 % of all species-level diversity among extant turtles (Iverson 1992). Geoemydidae is a clade of predominantly fresh-water aquatic, semiaquatic, and terrestrial turtles, but its habitats also include coastal marine areas and tropical forests. This "Old World" clade is widely distributed in Europe, North Africa, India, southern Russia, and the Philippines; only the genus *Rhinoclemmys* is present in Central America and Northern South America (Iverson 1992) (Fig. 2.3a).

In agreement with Hirayama (1984), extinct and extant species of Geoemydidae (Batagurinae, according to his study) are characterized by a distinct musk duct foramina enclosed within the third and seventh peripheral carapace bones. Several authors (i.e., McDowell 1964; Bramble 1974; Bickham 1975; Hirayama 1984; Carr and Bickham 1986; Yasukawa et al. 2001; Honda et al. 2002a, b; Spinks et al. 2004; Le and McCord 2008) have contributed with morphological and molecular data to enhance our knowledge of the systematic of this clade. McDowell (1964) divided Emydidae in two clades, Emydinae and Batagurinae (a junior synonymous of Geoemydidae *sensu* Bour and Dubois 1986). Hirayama (1984) presented the first phylogenetic analyses of Geoemydidae and proposed the extinct taxon *Echmatemys* as the sister taxon of remaining Geoemydidae, as well as a sister-taxa relationship between the "broad-jawed" and "narrow-jawed" lineages. The formed group includes the highly aquatic *Batagur*, *Hardella*, and *Orlitia* complexes, while the latter is composed of terrestrial turtles with a less extensive secondary palate, including members of the *Geoemyda* complex (sensu, McDowell 1964) and the tortoises (Testudinidae). A different phylogenetic scenario is proposed by Spinks et al. (2004) based on maximum parsimony, maximum likelihood, and Bayesian analysis of 65 species using mitochondrial and nuclear DNA. Spinks et al. (2004) recognized three main lineages, the Testudinidae, the Old World Geoemydidae, and the Central and South American genus *Rhinoclemmys*.

Paraphyletic relationships between Geoemydidae and Testudinidae have been suggested by morphological (McDowell 1963; Hirayama 1984) and molecular studies (Lamb and Lydeard 1994; Spinks et al. 2004). However, other contributions seem to support the monophyly of this clade (Claude and Tong 2004; Le and McCord 2008).

Genera included: *Batagur* Gray 1856b, *Bergouniouxchelys* Hervet 2004a; Borkenia Schleich 1994, *Clemmydopsis* Boda 1927, *Cuora* Gray 1856a, *Cyclemys* Bell 1834, *Cucullemys* Hervet 2004a, *Cuvierichelys* Hervet 2004a, *Echmatemys* Hay 1908, *Epiemys* Yeh 1963, *Francellia* Hervet 2004a, *Euroemys* Hervet 2004a, *Geiselemys* Khosatzky and Mlynarski 1966, *Geoclemys* Gray 1856b, *Geoemyda* Gray 1834, *Guandongemys* Claude, Zahng, Li, Mo and Tong 2012, *Hardella* Gray 1870, *Heosemys* Stejneger 1902, *Hummelemys* Hervet 2004b, *Juvemys* Hervet 2003, *Landreatchelys* Hervet 2004a, *Leucocephalon* McCord, Iverson, Spinks and Shaffer 2000, *Malayemys* Lindholm 1931, *Mauremys* Gray 1869a, *Melanochelys*

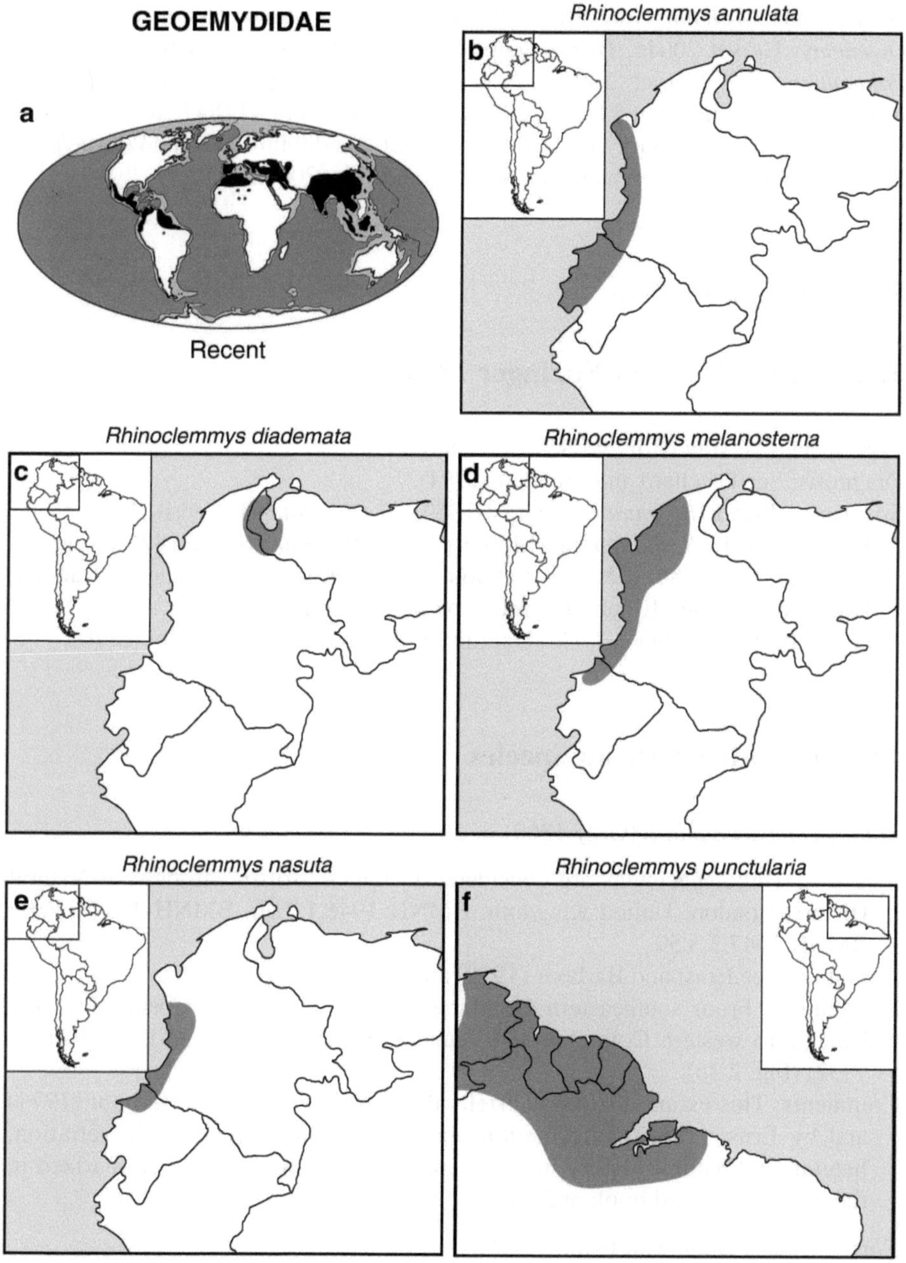

Fig. 2.3 Geographic distribution of Geoemydidae in recent times. (**a**) Global distribution of Geoemydidae. (**b**) Geographic distribution of *Rhinoclemmys annulata* in South America. (**c**) Geographic distribution of *Rhinoclemmys diademata* in South America. (**d**) Geographic distribution of *Rhinoclemmys melanosterna* in South America. (**e**) Geographic distribution of *Rhinoclemmys nasuta* in South America. (**f**) Geographic distribution of *Rhinoclemmys punctularia* in South America

Gray 1869b, *Morenia* Gray 1870, *Notochelys* Gray 1863b, *Orlitia* Gray 1873, *Owenemys* Hervet 2004a, *Palaeochelys* Meyer 1847, *Palaeoemys* Schleich 1994, *Palaeomauremys* Hervet 2004a, *Pangshura* Gray 1856b, *Promalacoclemmys* Reinach 1900, *Provencemys* Hervet 2004a, *Ptychogaster* Pomel 1847, *Rhinoclemmys* Fitzinger 1836, *Sacalia* Gray 1870, *Siebenrockiella* Lindholm 1929, and *Vijayachelys* Praschag, Schmidt, Fritzsch, Müller, Gemel, and Fritz 2006.

South American Taxa

2.2.1 *Rhinoclemmys* Fitzinger 1836

Type species: *Testudo dorsata* Schoepff 1801.
Diagnosis: See Pritchard and Trebbau (1984).
Species included: *R. annulata* (Gray 1860), *R. areolata* (Duméril, Bibron and Duméril 1851), *R. diademata* (Mertens 1954), *R. funera* (Cope 1876), *R. melanosterna* (Gray 1861), *R. nasuta* (Boulenger 1902), *R. panamaensis* (Cadena, Bourque, Rincon, Bloch, Jaramillo and MacFadden 2012), *R. pulcherrima* (Gray 1956b), *R. punctularia* (Daudin 1801), and *R. rubida* (Cope 1870).

Extant South American Species

Rhinoclemmys annulata (Gray 1860)

Syntypes: Three specimens are considered syntypes, British Museum of Natural History, London, United Kingdom. BMNH 1946.1.2.56, BMNH 1947.3.5.58, BMNH 1947.3.5.59.
Definition: See Ernst and Barbour (1979) and Ernst (1980).
Distribution: From southeastern Honduras, eastern Nicaragua, Costa Rica, and Panama to western Colombia and Ecuador (Ernst and Barbour 1979; Iverson 1992) (Fig. 2.3b).
Comments: This extant turtle was briefly described by Ernst and Barbour (1979) and by Ernst (1980). Information about distribution, geographic variation, habitat, and natural history of this terrestrial Geoemydidae is summarized in the aforementioned book and paper.

Rhinoclemmys diademata (Mertens 1954)

Holotype: Senckenberg Museum Frankfurt, Frankfurt, Germany. SMF 48141.
Diagnosis: See Pritchard and Trebbau (1984).
Distribution: From Western Zulia State in Venezuela, the adjacent parts of Tachira and Trujillo Departments, and Norte de Santander Department in Colombia (Pritchard and Trebbau 1984) (Fig. 2.3c).

Comments: A detailed morphological description of this species, originally named as *Geoemyda punctularia diademata* by Mertens (1954), was published by Pritchard and Trebbau (1984). Additional data about size and growth, geographic variation, habitat, feeding, and reproduction was also described by the authors mentioned above.

Rhinoclemmys melanosterna (Gray 1861)

Syntypes: Two specimens of the British Museum of Natural History, London, United Kingdom. BMNH 1947.3.4.8 and BMNH 1947.

Definition: See Ernst and Barbour (1989).

Distribution: Eastern Panama to northwestern Colombia and extreme northwestern Ecuador (Iverson 1992) (Fig 2.3d).

Comments: This species was originally described as *Geoclemys melanosterna* by Gray (1861) and restricted to a variant of *punctularia* by Boulenger in 1889. Pritchard (1979) proposed the species status of *melanosterna* and the taxonomic designation was followed by Pritchard and Trebbau (1984), Paolillo (1985), and Carr and Bickham (1986). Additional information about the geographic variation, habitat, and natural history of this species is reported by Medem (1962), Castaño Mora and Medem (1983), and Ernst and Barbour (1989).

Rhinoclemmys nasuta (Boulenger 1902)

Types: British Museum of Natural History, London, United Kingdom. Four syntypes BMNH 1947.3.5.54, BMNH 1947.3.5.55, BMNH 1947.3.5.56, and BMNH 1947. 3.5.57.

Definition: See Ernst and Barbour (1979). See also a morphological description of this species in Carr and Girardo (2009).

Distribution: Caribbean-drained middle and upper Río Atrato basin of northwestern Colombia and from the Pacific Ocean coastal drainages from the Río Baudó in Colombia south to the Río Esmeraldas in Ecuador (Carr and Girardo 2009 and references therein) (Fig. 2.3e).

Comments: Additional data concerning to the taxonomy, habitat, ecology, population status, and conservation of the most aquatic species of the genus *Rhinoclemmys* were summarized by Carr and Girardo (2009).

Rhinoclemmys punctularia (Daudin 1801) (Fig. 2.4)

Holotype: Muséum National d'Histoire Naturelle, Paris, France. MNHN 9130.

Diagnosis: See Pritchard and Trebbau (1984), authors that also provided a morphological description of this species.

Distribution: This species is distributed in eastern Venezuela, Trinidad, and into northeastern Brazil and probably is widespread in all the three Guianas (Pritchard and Trebbau 1984) (Fig. 2.3f).

Comments: This species was originally named *Testudo punctularia* by Daudin (1801) and later referred to *Geoemyda* by Wemuth and Mertens (1961) and Pritchard (1964, 1967). The complete taxonomic history of this species is

Fig. 2.4 Drawing of *Rhinoclemmys punctularia*

detailed in Fretey et al. (1977) who also redescribed Daudin's (1801) type. Pritchard and Trebbau (1984) summarized the available information concerning size and growth, geographic variation, habitat, feedings, reproduction, economic importance, and vernacular names of this species.

2.3 Chelydridae Gray 1831a

Presently, this is a semiaquatic New World clade, distributed throughout the Americas, from southern Canada to Ecuador (Fig. 2.5a). Two extant genera are recognized: *Chelydra* Schweigger (1814) and *Macrochelys* Gray 1856a (Fritz and Havas 2007). The species of the former genus (*Chelydra*) is widespread in the Americas, whereas the latter (*Macrochelys* Gray 1856a) is restricted to southeastern North America (see Ernst and Barbour 1979; Fritz and Havas 2007). Extinct species are recognized from the Palaeocene to the Pleistocene in North America and from the Oligocene to the Pliocene of Europe and Asia (Broin 1977; Lapparent de Broin 2001). The snapping turtles have big skulls with emarginated temporal regions and lack a secondary palate. The carapace is attached to the reduced, cross-shaped plastron by a narrow bridge. The carapace is keeled and exhibits costo-peripheral fontanellae in adult specimens. The entoplastron is T-shaped with a median plastral fontanella. A detailed diagnosis of the Chelydridae based on skull and lower jaw was provided by Gaffney (1975). Phylogenetic studies based on morphology (e.g., Gaffney 1975; Shaffer et al. 1997) suggest a close relationship between Chelydridae, *Platysternon* Gray 1831b and *Macrocephalochelys* Pidoplichko and Tarashchuk 1960, a hypothesis that has been rejected by genetic data (Krenz et al. 2005; Parhan et al. 2006). The relationships among *Chelydra* species have been addressed by Phillipps et al. (1996).

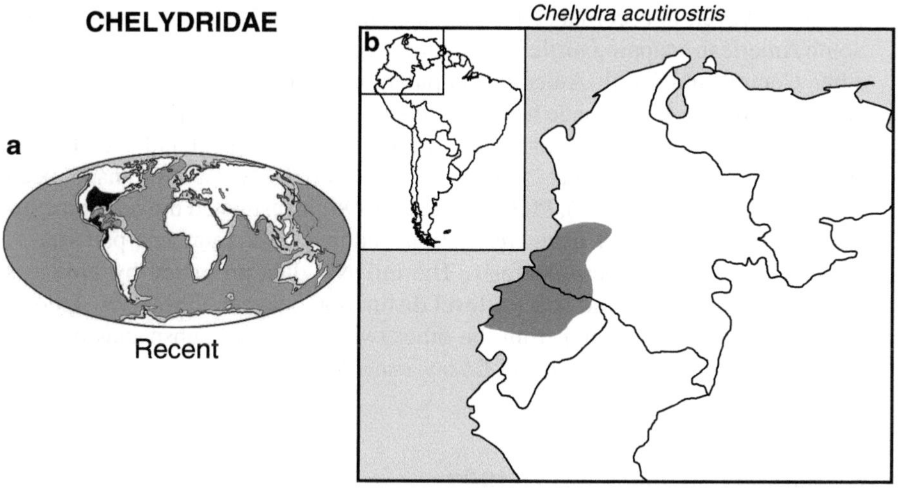

Fig. 2.5 Geographic distribution of Chelydridae in recent times. (**a**) Global distribution of Chelydridae. (**b**) Geographic distribution of *Chelydra acutirostris* in South America

Genera included: *Acherontemys* Hay 1899, *Chelydra* Schweigger 1814, *Chelydrops* Matthew 1924, *Chelydropsis* Peters 1868, *Denverus* Hutchison and Holroyd 2003, *Emarginochelys* Whetstone 1978, *Macrochelys* Gray 1856a, *Protochelydra* Erickson 1973, and *Tullochelys* Hutchison 2013.

South American Taxa

2.3.1 *Chelydra* Schweigger 1812

Type species: *Testudo serpentina* Linnaeus 1758.
Diagnosis: Pritchard (1979).
Species included: *Chelydra acutirostris* Peters 1862, *Chelydra rossignonii* (Bocourt 1868), and *Chelydra serpentina* (Linnaeus 1758).

Chelydra acutirostris Peters 1862

Holotype: Naturhistoriska Riksmuseet, Stockholm, Sweden. GA 49.
Diagnosis: See Pritchard (1979).
Distribution: Southern Central America and northern South America, from southern Honduras to the Pacific Ocean lowlands of Colombia and Ecuador (Fritz and Havas 2007) (Fig. 2.5b).
Comments: Traditionally, the South American species of *Chelydra* was assigned to *Chelydra serpentina acutirostris*, a subspecies of *Chelydra serpentina* (Pritchard

1979). However, in accordance to Fritz and Havas (2007) classification list, the South American snapping turtle is recognized as a different species, distinct from other Central and North American populations of *Chelydra serpentina*. This taxonomic decision was made based on genetic studies by Phillipps et al. (1996). Since previous phylogenetic analyses based on morphological and osteological characters including distinct subspecies of *Chelydra serpentina* have been inconclusive (Medem 1977), Phillipps et al. (1996) investigated the relationships among these subspecies using restriction endonuclease fragment patterns of mtDNA and protein electrophoresis. The mtDNA data presented by Phillipps et al. (1996) supported the species-level distinction between *Che. s. rossignonii* and *Che. s. acutirostris* and from the other two recognized subspecies of *Che. serpentina*: *Che. s. serpentina* and *Che. s. osceola*.

2.4 Kinosternidae Agassiz 1857

The American mud or musk turtles constitute a New World clade of small aquatic to semiaquatic species distributed from Canada to northern and central South America (Fig. 2.6a). These turtles evolved in the Americas, and extinct species are known from the Oligocene of South Dakota (Hay 1906; Williams 1952). Today, this clade is represented by 4 genera and 24 species (Rhodin et al. 2010) with the highest diversity in Mexico and Central America (Smith and Smith 1979). According to Ernst and Barbour (1989), Kinosternidae is characterized by elongated skulls with moderately emarginated temporal region, reduced frontal bones excluded from the margin of the orbits, maxilla–quadratojugal contact, separate squamosal and postorbital, secondary palate, carapaces that range from flattened to slightly domed with one to three longitudinal keels, a riblike lateral process on the nuchal bone, 10 pairs of peripheral bones and 22 marginal scales plus one cervical, a plastron with one or two hinges, a variable number of plastral scutes (from 7 or 8 to 10 or 11), musk glands associated with the bridge, only one biconvex cervical vertebra, and procoelous caudal vertebrae.

Phylogenetic relationships among extant species based on protein variations and morphology have been discussed by several authors (Seidel et al. 1986; Iverson 1988, 1991; Hutchison 1991). The homology of the plastral scales was discussed by Hutchison and Bramble (1981), and the structure and evolution of shell kinesis was analyzed by Bramble et al. (1984). Typically, two clades are recognized: Kinosterninae and Staurotypinae (see Ernst and Barbour 1989; Iverson 1992). Although the former was elevated to family status on the basis of chromosomal differences from Kinosterninae reported by Bickham and Carr (1983), this decision is not supported by morphological studies (Iverson 1988, 1991; Hutchison 1991; Gaffney and Meylan 1988; Meylan and Gaffney 1989).

Genera included: *Baltemys* Hutchison 1991, *Claudius* Cope 1865, *Kinosternon* Spix 1824, *Staurotypus* Wagler 1830, *Sternotherus* Bell in Gray 1825, and *Xenochelys* Hay 1906.

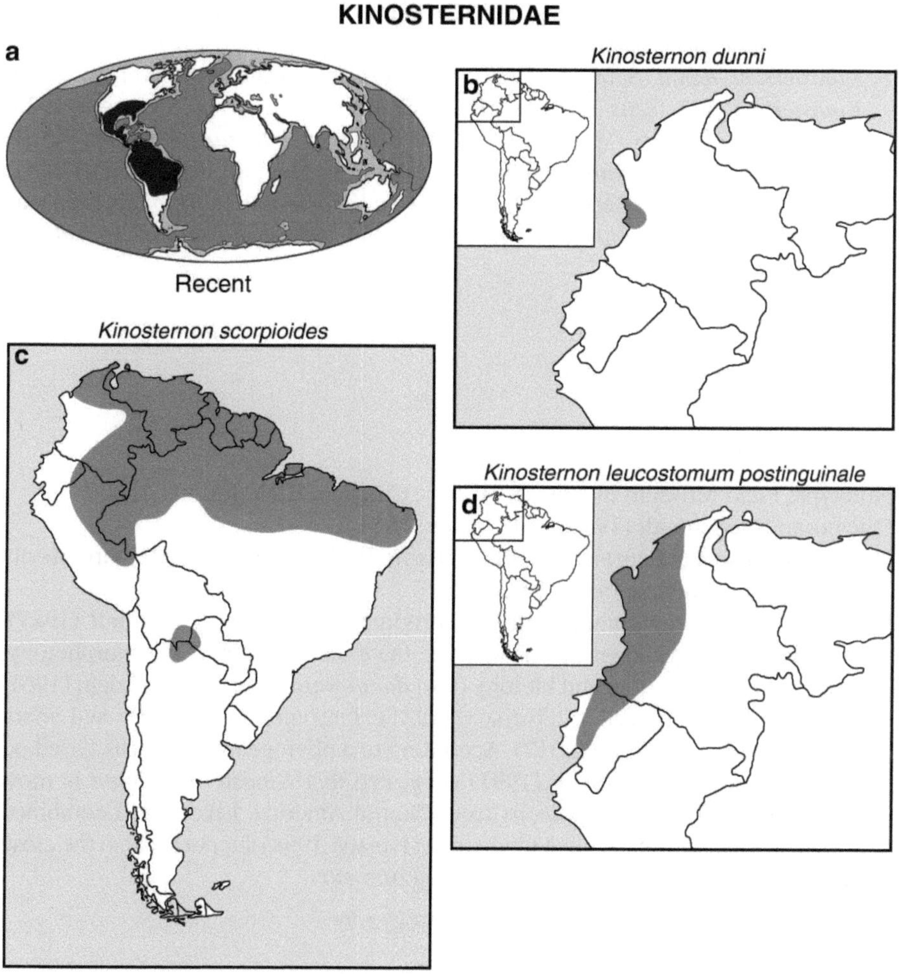

Fig. 2.6 Geographic distribution of Kinosternidae in recent times. (**a**) Global distribution of Kinosternidae. (**b**) Geographic distribution of *Kinosternon dunni* in South America. (**c**) Geographic distribution of *Kinosternon scorpioides* in South America. (**d**) Geographic distribution of *Kinosternon leucostomum postinguinale* in South America

South American Taxa

2.4.1 Kinosternon *Spix 1824*

Type species: *Kinosternon longicaudatum* Spix 1824.

Species included: *Kinosternon acutum* Gray 1831c, *Kinosternon alamosae* Berry and Legler 1980, *Kinosternon angustipons* Legler 1965, *Kinosternon arizonense*

Gilmore 1922, *Kinosternon baurii* Garman 1891, *Kinosternon chimalhuaca* Berry, Seidel and Iverson in Rogner 1996, *Kinosternon creaseri* Hartweg 1934, *Kinosternon dunni* Schmidt 1947, *Kinosternon durangoense* Iverson 1979, *Kinosternon flavescens* Agassiz 1857, *Kinosternon herrerai* Stejneger 1925, *Kinosternon hirtipes* Wagler 1830, *Kinosternon integrum* LeConte 1854, *Kinosternon leucostomum* Duméril and Bibron in Duméril and Duméril 1851, *Kinosternon oaxacae* Berry and Iverson 1980, *Kinosternon scorpioides* Linnaeus 1766, *Kinosternon sonoriense* LeConte 1854, and *Kinosternon subrubrum* Bonnaterre 1789.

South American Species

Kinosternon dunni Schmidt 1947

Holotype: Field Museum of Natural History, Chicago, USA. FMNH 42804.
Diagnosis: See Schmidt (1947) and Iverson (1981).
Distribution: Río San Juan and Río Baudo basins, Pacific Ocean coastal environments of Colombia (Iverson 1992) (Fig. 2.6b).
Comments: The Dunn's mud turtle was originally described by Schmidt (1947). Although little is known of this species, the data available of the morphology, geographic distribution, and biology of *K. dunni* were reported by Medem (1961, 1962) and summarized by Iverson (1981), Groombridge (1982), and more recently by Iverson et al. (2012). According to a phylogenetic analysis based on morphological data, Iverson (1991) suggested that *Kinosternon dunni* is most closely related to *K. angustipons* from Central America. Likewise, a combined molecular and morphological phylogeny (Iverson 1998) supported also the close relationship between *K. dunni* and K. *leucostomum*.

Kinosternon scorpioides (Linnaeus 1766) (Fig. 2.7)

Holotype: Not located.
Diagnosis: See Pritchard and Trebbau (1984).
Distribution: From Southern Mexico to Brazil and northern Argentina (Iverson 1992) (Fig. 2.6c).
Comments: Six subspecies of the scorpion mud turtle have been recognized, but only three are found in South America (Iverson 1992; Rhodin et al. 2010). The South American *Kinosternon scorpioides* includes the subspecies: *K. s. scorpioides* (Linnaeus 1766), *K. s. seriei* Freiberg 1936, and *K. s. carajasensis* Cunha 1970. The nominal form is widespread in northern South America (Pritchard and Trebbau 1984), whereas *K. s. seriei* and *K. s. carajasensis* are more restricted in their ranges. The former occurs in the Gran Chaco (Argentina), and the latter inhabits in the Serras dos Carajás (State of Pará, Brazil). Since the two named subspecies (*K. s. seriei* and *K. s. carajasensis*) were described on the basis of inadequate samples, Cabrera and Colantonio (1997) discussed the taxonomic

Fig. 2.7 Drawing of *Kinosternon scorpioides*

significance of their suggested allopatry and proposed to include *K. s. seriei* and *K. s. carajasensis* in synonymy with *K. s. scorpioides*. Data regarding the distribution, habitat, ecology, and economic importance of *K. s. scorpioides* is summarized by Pritchard and Trebbau (1984).

Kinosternon leucostomum Duméril and Bibron in Duméril and Duméril 1851

Syntypes: Muséum National d´Histoire Naturelle de Paris, France. MNHN 21114, MNHN 8311, MNHN 9088. The specimen MNHM 8211 was designed lectotype of this species by Stuart (1963).
Definition: See Berry (1978) and Ernst and Barbour (1989).
Distribution: Mexico through Central America to western Colombia, western Ecuador, and probably northern Peru (Iverson 1992).

Kinosternon leucostomum postinguinale Cope 1887

Syntypes: National Museum of Natural History, Smithsonian Institution, Washington, USA. USNM 45582, USNM 51165.
Definition: See Berry (1978) and Ernst and Barbour (1989).
Distribution: Range from Nicaragua to Ecuador or northern Peru (Iverson 1992) (Fig. 2.6d).
Comments: *Kinosternon leucostomum* was reviewed by Berry (1978) and Smith and Smith (1979). Two subspecies were recognized by Berry (1978): the northern white-lipped mud turtle (*K. leucostomum leucostomum*) and the southern white-lipped mud turtle (*K. leucostomum postinguinale*). The latter subspecies is the only one known in South America, and contrarily to the northern subspecies, it is characterized by a flattened carapace, a narrow plastron, a short gular scale, and the inguinal setback on the bridge (Ernst and Barbour 1989).

2.5 Trionychidae Fitzinger 1826

Trionychidae, or softshell turtles, is a highly specialized clade of aquatic Testudines characterized by rounded and flattened shells with an extreme reduction of dermal bones that are covered by a leathery skin rather than horny scales, paddle-like limbs with a significant reduction in the number of claws, and carnivorous habits. Extant softshell turtles are found in North America, Asia, Africa, and the Indo-Australian archipelago (Fig. 2.8a); however, their fossil record indicates that they were more widespread in the past, with occurrences in Europe (Broin 1977; Lapparent de Broin 2001) and in northern South America (Wood and Patterson 1973; Sánchez Villagra et al. 2004; Head et al. 2006). As noted by Head et al. (2006), fossil softshells are mostly absent from southern continents, and the poor record is the result of distinct dispersal events from North America and Asia. Head et al. (2006) summarized the trionychid fossil record of Gondwana landmass, indicating their presence in Africa during the Late Neogene (immigrants from South Asia, Wood 1979; Meylan et al. 1990), during the Eocene and Pliocene in Australia (White 1998; Gaffney and Bartholomai 1979), absent from Indian prior to the connection with Asia in the Paleogene (Broin 1987; Head et al. 1999), and in northern South America during the Miocene–Pliocene (see references above).

Genera included: *Altatrionyxs* Chkhikvadze 2008, *Amyda* Geoffroy Saint-Hilaire 1809, *Apalone* Rafinesque 1832, *Aspideretes* Hay 1904, *Aspideretoides* Gardner, Russell and Brinkman 1995, *Atoposemys* Hutchison 2013, *Axestemys* Hay 1908, *Chitra* Gray 1844, *Cyclanorbis* Gray 1854, *Cycloderma* Peters 1854, *Dogania* Gray 1844, *Francedebroinella* Chkhikvadze 1999, *Gilmoremys* Joyce and Lyson 2011, *Helopanoplia* Hay 1908, *Hutchemys* Joyce, Revan and Danilov 2009, *Khunnuchelys* Brinkman, Nessov and Peng 1993, *Lissemys* Smith 1931, *Nilssonia* Gray 1872,

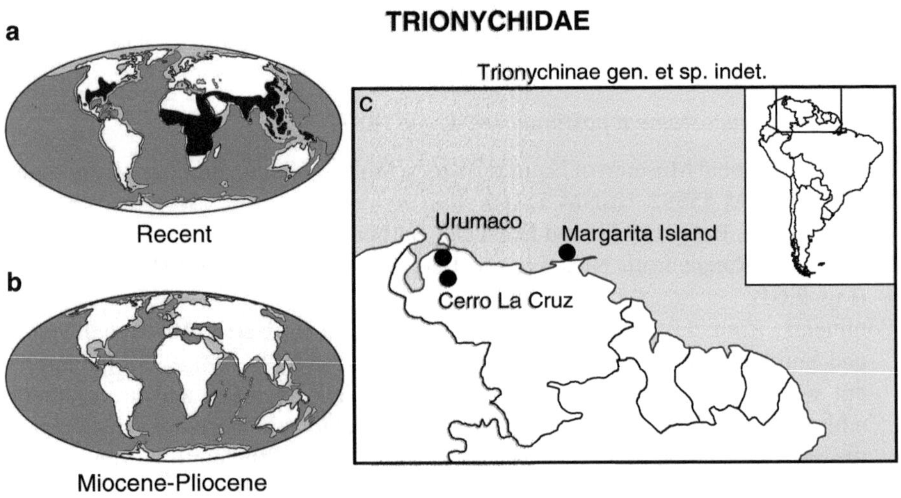

Fig. 2.8 Geographic distribution of Trionychidae in recent times. (**a**) Global distribution of Trionychidae. (**b**) Paleogeographic reconstruction of continents during the Miocene–Pliocene. (**c**) Geographic and stratigraphic location of the South American record of trionychins

Palaeotrionyx Schmidt 1945, *Palea* Meylan 1987, *Paraplastotomenus* Kordikova 1994a, *Pelochelys* Gray 1864, *Pelodiscus* Fitzinger 1835, *Rafetus* Gray 1864, *Sinamyda* Chkhikvadze 2000, *Trionyx* Geoffroy Saint-Hilaire 1809, and *Ulutrionyx* Kordikova 1994b.

South American Species

2.5.1 Trionychinae Fitzinger 1826

2.5.1.1 Gen. et sp. indet

Specimens: Museo de Ciencias Naturales Caracas, Caracas, Venezuela: MCNC 238, a hyoplastron. Museo de Biología, Universidad de Zulia, Maracaibo, Venezuela: MBLUZ-P 5046, ten carapace and plastron fragments. Universidad Nacional Experimental Francisco Miranda, Venezuela: UNEFM 102, a left hypoplastron.

Localities, horizons, and ages: The former specimen (MCNC 238) was discovered 12 km WSW of Urumaco (Falcon State, Venezuela) from the upper part of the Urumaco Formation (Wood and Patterson 1973). The specimen MBLUZ-P 5046 was recovered at Cerro La Cruz locality from sediments of the Castillo Formation, which is Early Miocene in age (Sánchez Villagra et al. 2004 and references therein). The last specimen (UNEFM 102) was found in locality PPP2573 (Head et al. 2006) on Margarita Island, Venezuela, from the La Tejita Member of the Cubagua Formation, which is Miocene–Early Pliocene in age (Aguilera and Rodrigues de Aguilera 2001) (Fig. 2.8b, c).

Comments: Wood and Patterson (1973) pointed out that the earliest, isolated fragment of trionychid that was discovered in northern South America from the Urumaco Formation represented failed waif immigration. However, the most recent discoveries of trionychids in an Early Miocene locality 100 km south to Urumaco (Sánchez Villagra et al. 2004) and in the Miocene–Early Pliocene of Margarita Island prompted Head et al. (2006) to suggest that trionychids might have successfully colonized fluvial and coastal regions of northern South America. The geographical restriction of South American trionychids to coastal and pericoastal areas along the northern part of this continent and the absence of trionychids in the remaining continental Neogene of South American (where chelid and pelomedusoids fossil remains are preserved) (e.g., Wood 1997; Cione et al. 2000), led Head et al. (2006) to consider the hypothesis of competitive exclusion of trionychids by pleurodires (Simpson 1943; Wood and Patterson 1973).

2.6 Discussion

Considering that the existence of terrestrial routes is thought to be a requisite for dispersion of aquatic continental turtles (see Pritchard 1984; Pritchard and Trebbau 1984), the timing of the arrival of aquatic continental cryptodiran species of different clades (Emydidae, Chelydridae, Geoemydidae, Kinosternidae, Trionychidae) in South America is a major issue.

Despite general agreement that Emydidae, and particularly *Trachemys*, were differentiated in North America and invaded South America more recently (Simpson 1943; Savage 1966; Pritchard 1984; Pritchard and Trebbau 1984; Seidel and Jackson 1990), disagreement remains regarding the timing of colonization. According to Vanzolini and Heyer (1985), Vanzolini (1995), and Seidel and Jackson (1990), *Trachemys* entered South America following the closure of the Panamanian Isthmus. To the contrary, Fritz et al. (2012) suggested that at least some species of *Trachemys* might have entered South America before the closure of the isthmus. Vanzolini and Heyer (1985) and Vanzolini (1995) argued that *Trachemys* colonized South America during the Early Pliocene based on the extremely disjointed range of South American emydid species. The arrival of *Trachemys* to the Neotropical Region and the aforementioned geographic disjointed distribution are interpreted by Vanzolini and Heyer (1985) as the result of an extensive dispersive event followed by extinction. According to these authors, the extinction of northern South American *Trachemys* species could be the result of competition among these species and the successful podocnemidid and chelid turtles. Vanzolini and Heyer's hypothesis lacks empirical support since there is no evidence of the species extinctions they suggested. Contrarily to this hypothesis, Seidel and Jackson (1990), in their review of the genus *Trachemys*, suggested that it radiated during the Miocene and Pliocene and entered the Neotropics during the Early Pleistocene after the closure of the isthmus. These ideas are based on Moll and Legler (1971) studies comparing the life cycle of *T. scripta* (*Pseudemys* in the cited paper) in Panama with those of other tropical turtles. The study suggests that this species is still in the process of adapting to a tropical environment. According to Moll and Legler (1971), the lack of adaptations to a tropical environment supports a relatively recent (Pleistocene) arrival of *Trachemys scripta* to Central America from North America. Vanzolini (1995) criticized Moll and Legler's (1971) argument, citing two main fallacies. According to Vanzolini (1995), the first fallacy is to judge adaptation relative to a preconceived model. The second fallacy lies in Moll and Legler (1971) model itself. Nonetheless, results of recent molecular phylogenies for Central and South American slider turtles suggest that South America was invaded by these turtles from Central America at least twice (Fritz et al. 2012). In this context the first lineage to have colonized South America was the clade *T. dorbigni* + *T. adiutrix* (as a subspecies of *T. dorbigni* in Fritz et al. 2012) prior to the emergence of the Panamanian land bridge, whereas *T. callirostris* (as a subspecies of *T. ornata* in Fritz et al. 2012 article) represents a more recent colonization occurring after the closure of the Panamanian Isthmus.

Recent paleontological findings in Central and South America and new phylogenetic and biogeographical analyses of Emydidae and Geoemydidae allowed several authors (i.e., Sánchez Villagra et al. 2004; Head et al. 2006; Le and McCord 2008; Cadena 2009; Fritz et al. 2012) to present more information supporting the dispersal events of some groups of aquatic continental cryptodires from Central to South America as early as the Early Miocene, prior to the formation of the Panama Isthmus (see Woodburne 2010). Regarding the new paleontological findings, two different

Miocene chelonian faunas from the Panamá Canal basin were described by Cadena (2009) and Cadena et al. (2012). These discoveries illuminate the history of colonization and faunal interchange between North, Central, and South America. In the former study (Cadena 2009), the fauna was recovered from the Early Miocene Culebra Formation (including trionychids and podocnemidids), whereas in the latter the fauna was discovered in the Early–Middle Miocene Cucaracha Formation (including a new species of the geoemydid *Rhinoclemmys*, a new species of the kinosternid *Staurotypus*, trionychids, and testudinids). According to Cadena (2009), the occurrence of a species of *Rhinoclemmys* and another of *Staurotypus* in the Early–Middle Miocene of Panama Canal basin is significant regarding the early dispersion of these groups of aquatic cryptodires in Central America. In this sense, the former not only represents the oldest record of Geoemydidae in Central America but also supports the presence of this genus very close to South America at this time, whereas the latter represents the earliest record of a mud turtle in Central America and indicates that this genus was also very close to South America at this time.

The other source of data is biogeographical patterns among *Rhinoclemmys* and *Trachemys* species (Le and McCord 2008; Fritz et al. 2012). Although alternative hypotheses regarding the dispersal of *Rhinoclemmys* into South America have been proposed (Duellman 1979; Vanzolini and Heyer 1985; Carr 1991), Carr's (1991) hypothesis that some species colonized South America prior to the closure of the Panama Isthmus is supported by Le and McCord (2008) studies. According to Le and McCord (2008), there is a correlation between some vicariant events and the radiation of *Rhinoclemmys* in Central and South America. Among the events, Le and McCord recognize different radiations related to the emergence of the Sierra Madres of Mexico (Oligocene), the Nuclear Highland (Miocene), and the Panama land bridge (Pliocene). The first vicariant event took place during the Early Miocene in Central America and might have isolated the *R. areolata* group in the Atlantic Ocean lowlands from *R. rubida* in the Pacific Ocean lowlands. The second event took place during the Middle Miocene with the uplift of the Nuclear Highland, and it might have caused the divergence of *R. pulcherrima* and *R. annulata* and the isolation of *R. areolata* from the *R. punctularia* and *R. funerea* groups. In agreement with the Le and McCord's phylogeny and with the distribution of the species (*R. punctularia* + *R. funerea*, *R. annulata* and *R. nasuta*), *Rhinoclemmys* species colonized South America at least four times, three of them might have occurred prior to the emergence of the Panamanian Isthmus (Le and McCord 2008, Fig. 6).

According to Woodburne (2010), the paleogeographic scenario for the Great American Biotic Interchange (GABI) well before final closure of the Central American Seaway is evident in the geological and tectonic reconstruction proposed by Coates et al. (2004). In this reconstruction, there are numerous islands in the Panamanian region as early as the Mio-Pliocene boundary (6 Ma). During the Middle Miocene (12 Ma), there were emergent lands in Mexico, Honduras, Nicaragua, and Colombia, with the Central American Seaway being largely interrupted by the evolving volcanic arc. This paleogeographic model and the suggested ability of these types of aquatic cryptodirans to cross sea straits (Fritz et al. 2006,

2007, 2008) led Le and McCord (2008) and Fritz et al. (2012) to propose that the ancestor of *T. dorbigni* and *T. adiutrix* and *Rhinoclemmys* arrived in South America prior to the emergence of Panamanian Isthmus via island hopping or transoceanic dispersal.

References

Agassiz L (1857) Contributions to the natural history of the United States of America, vol I. Little Brown and Co, Boston

Aguilera O, Rodrigues de Aguilera D (2001) An exceptional coastal upwelling fish assemblage in the Caribbean Neogene. J Paleontol 75:732–742

Alderton D (1988) Turtles and tortoises of the world. Fact on File Inc, New York

Barbour T, Carr AF (1940) Antillean terrapins. Mem Mus Comp Zool Harvard 54:381–415

Bell T (1834) Characters of a new genus of freshwater tortoise (*Cyclemys*). Proc Zool Soc Lond 1834:17

Berry JF (1978) Variation and systematics in the *Kinosternon scorpioides* and *K. leucostomum* complexes (Reptilia: Testudines: Kinosternidae) of Mexico and Central America. Doctoral dissertation, University of Utah, Salt Lake City

Berry JF, Iverson JB (1980) A new species of mud turtle genus *Kinosternon* from Oaxaca, Mexico. J Herpetol 14:313–320

Berry JF, Legler JM (1980) A new turtle (genus *Kinosternon*) from Sonora Mexico. Contrib Sci Nat Hist Mus Los Angel Co 325:1–12

Bickham JW (1975) A cytosystematic study of turtles in the genera *Clemmys*, *Mauremys* and *Sacalia*. Herpetologica 31:198–201

Bickham JW, Carr JL (1983) Taxonomy and phylogeny of the higher categories of cryptodiran turtles based on a cladistic analysis of chromosomal data. Copeia 1983:918–932

Bickham JW, Lamb T, Minx P, Patton JC (1996) Molecular systematic of the genus *Clemmys* and the intergeneric relationships of emydid turtles. Herpetologica 52:89–97

Bock BC, Páez VP, Daza JM (2010) *Trachemys callirostris* (Gray 1856) Colombian Slider, Jicotea, Hicotea, Galápago, Morrocoy de Agua. In: Rhodin AGJ, Pritchard PCH, van Djik PP, Saumure RA, Buhlman KA, Iverson JB, Mittermeier RA (eds) Conservation biology of freshwater turtles and tortoises: a compilation project of the IUCN/SSC tortoises and freshwater turtle specialist group, vol 5, Chelonian research monographs. Chelonian Research Foundation, Lunenburg, pp 042.1–042.9

Bocourt F (1868) Description de quelques chélonniens nouveaux appartenant a la faune Mexicaine. Ann Sci Nat Zool Paris 10:121–122

Boda A (1927) *Clemydropsis sopronensis*, n.g.n.sp. aus der unteren pannonischen Stufe von Sopron in Ungarn. Centr Min Geol Palaont 9:375–383

Bonnaterre PJ (1789) Tableau Encyclopédique et Méthodique des Trois Règnes de la Nature. Erpétologie. Panckoucke, Hôtel deThou, Paris

Boulenger GA (1889) Catalogue of the Chclonians. Rhynchocephalians, and Crocodiles in the British Museum (Natural History). Taylor & Francis, London

Boulenger GA (1902) Descriptions of new batrachians and reptiles from northwestern Ecuador. Ann Mag Nat Hist Lond 9:51–57

Bour R, Dubois A (1986) Nomenclature ordinale et familiale des tortues (Reptilia). Note complimentaire. Bull Mens Linn Lyon 55:87–90

Bramble DM (1974) Emydid shell kinesis: biomechanics and evolution. Copeia 1974:707–727

Bramble DM, Hutchison JH, Legler JM (1984) Kinosternid shell kinesis: structure, function and evolution. Copeia 1984:456–475

Brinkman DB, Nessov LA, Peng JH (1993) *Khunnuchelys* gen. nov., a new trionychid (Testudines: Trionychidae) from the Late Cretaceous of Inner Mongolia and Uzbekistan. C J Earth Sci 30:2214–2223

Cabrera MR (1995) Comparative composition of turtles species in four natural region of the Chacoan Domain, South America. An Mus Hist Nat Valparaíso 23:41–52, Chile
Cabrera MR, Colantonio SE (1997) Taxonomic revision of the South American subspecies of the turtle *Kinosternon scorpioides*. J Herpetol 31:507–513
Cabrera MR, Luna CA (2011) Presencia de *Trachemys* Agassiz, 1857 (Testudines, Emydidae) en el Pleistoceno tardío del centro de la Argentina. Cuad Herpetol 25:36–41
Cadena E (2009) New Cretaceous and Cenozoic Fossil turtles from Colombia and Panama; systematic paleontology, phylogenetical and paleobiogeographical implications. Dissertation MSc, University of Florida, Gainesville
Cadena E, Bourque JR, Rincon A, Bloch JI, Jaramillo CA, MacFadden BI (2012) New turtles (Chelonia) from the Late Eocene through Late Miocene of the Canal Basin. J Paleontol 86:539–557
Carr JL (1991) Phylogenetic analysis of the Neotropical turtle genus *Rhinoclemmys* Fitzinger (Testudines: Emydidae). Dissertation doctoral thesis, Southern Illinois University, Carbondale
Carr JL, Bickham JW (1986) Phylogenetic implications of karyotype variation in the Batagurinae (Testudines: Emydidae). Genetica 70:89–106
Carr JL, Girardo A (2009) *Rhinoclemmys nasuta* (Boulenger 1902) – large-nosed wood turtle, Chocoan river turtle. In: Rhodin AGJ, Pritchard PCH, van Dijk PP, Saumure RA, Buhlmann KA, Iverson JB, Mittermeier RA (eds) Conservation biology of freshwater turtles and tortoises: a compilation project of the IUCN/SSC tortoise and freshwater turtle specialist group, vol 5, Chelonian research monographs. Chelonian Research Foundation, Lunenburg, pp 034.1–034.6
Castaño Mora O, Medem F (1983) Datos preliminaries sobre la reproducción de *Rhinoclemmys melanosterna* Gray (Reptilia: Chelonia: Emydidae). Lozania 47:1–5
Chkhikvadze VM (1999) Some fossil soft-shell turtles of Asia (Rafetini trib. nov.). Trudy Tbilisskogo Gosudarstvennogo Pedagogicheskogo Universiteta 5:215–225 (In Russian)
Chkhikvadze VM (2000) On the systematic position of some extinct trionychids of North America and Asia. Trudy Tbilisskogo Gosudarstven 3:85–95 (In Russian)
Chkhikvadze VM (2008) Soft-shell turtles (Trionychidae) of Asia and North America: morphology, phylogeny, systematics, terminology of elements of carapace. Probl Paleobiol Tbilisi 3:85–95 (In Russian)
Cione AL, Azpelicueta M, Bond M, Carlini A, Casciotta R, Cozzuol M, de la Fuente M, Gasparini Z, Goin J, Noriega J, Scillato-Yané J, Soibelzon L, Tonni EP, Verzi D, Vucetich MG (2000) Miocene vertebrates from Paraná, eastern Argentina. Correlación Geol 14:191–237
Claude J, Tong H (2004) Early Eocene testudinoid turtles from Saint Papoul, France, with comments on the early evolution of modern Testudinoidea. Oryctos 5:3–45
Claude J, Zahng JY, Li JJ, Mo JY, Tong H (2012) Geoemydid turtles from the Late Eocene Maoning basin, southern China. Bull Soc Géol Fr 183:641–651
Coates AG, Colling LS, Aubry MP, Berggren WA (2004) The geology of the Darien, Panama, and the Miocene-Pliocene collision of the Panama arc with northwestern South America. Geol Soc Am Bull 116:1327–1344
Cope ED (1865) Third contribution to the herpetology of tropical America. Proc Acad Nat Sci Phila 17:185–198
Cope ED (1876) On the Batrachia and Reptilia of Costa Rica with notes on the herpetology and ichthyology of Nicaragua and Peru. 1. Acad Nat Sci Phila 8:93–154
Cope ED (1887) Catalogue of batrachians and reptiles of Central America and Mexico. Bull U S Natl Mus 32:1–98
da Cunha OR (1970) Uma nova subspécie de quelônio, *Kinosternon scorpiodes carajasensis* da Serra dos Carajás, Pará (Testudinata, Kinosternidae). Bol Mus Paraense Emilio Goeldi Zool 73:1–12
Daudin FM (1801) Histoire naturelle generale et particuliere des reptiles. Dufan Paris 2:1–432
de Broin F (1977) Contribution à l'étude des chéloniens. Chéloniens continentaux du Crétacé et du Tertiaire de France. Mém Mus Nat d'Hist Nat sér C 38:1–366

de Broin F (1987) Lower vertebrates from the early- middle Eocene Kuldana Formation of Kohat (Pakistan): Chelonia. Contrib Mus Paleont Univ Michigan 27:151–186

de la Fuente MS, Noriega J, Piña C (2002) *Trachemys dorbigni* (Dúmeril y Bibron, 1835) (Cryptodira: Emydidae) en el Pleistoceno Tardío de la Provincia de Entre Ríos, Argentina. Cuad Herpetol 16:65–72

Del Barco DM, Larriera A (1991) Sobre la validez de las subespecies de *Trachemys dorbigni* y su distribución geográfica. Rev Asoc Cienc Nat Litoral 22:11–17

Duellman WE (1979) The South American herpetofauna: a panoramic view. In: Duellman WE (ed) The South American herpetofauna: its origin, evolution, and dispersal. University of Kansas Printing Service, Kansas City

Duméril AMC (1806) Zoologie analytique au methode naturelle de classification des animaux. rendue plus facile a l'aide de tableaux synoptiques. Allais, Paris

Duméril AMC, Bibron G (1835) Erpetologie generale ou histoire naturelle complete des reptiles, vol 2. Librairie Encyclopedique de Roret, Paris

Duméril AMC, Duméril AHA (1851) Catalogue methodique de la collection des reptiles du Muséum d'Histoire Naturelle de Paris. Gide and Boudry, Paris

Erickson BR (1973) A new chelydrid turtle *Protochelydra zangerli* from the Late Paleocene of North America. Publ Sci Mus Minn 2:1–16

Ernst CH (1980) Rhinoclemmys annulata. Cat Am Amphib Rept 250:1–2

Ernst CH (2003) Trachemys callirostris. Cat Am Amphib Rept 768:1–4

Ernst CH, Barbour RW (1989) Turtles of the world. Smithsonian Institution Press, Washington, DC

Fitzinger LJ (1826) Neue Classification der Reptilien nach ihren Natürlichen Verwandtschaften nebst einer Verwandtschafts-Tafel und einem Verzeichnisse der Reptilien-Sammlung des Zoologischen Museum zu Wien. Hübner J G, Wien

Fitzinger LJ (1835) Entwurf einer systematischen Anordnung der Schildkröten nach den Grundsätzen der natürlichen Methode. Ann Nat Mus Wien 1:103–128

Freiberg M (1936) Una nueva Tortuga del norte argentine. Physis 12:169–171, Buenos Aires

Freiberg MA (1969) Una nueva subespecie de *Pseudemys dorbigni* (Dumeril et Bibron) (Reptilia, Chelonia, Emydidae). Physis 28:299–314, Buenos Aires

Fretey J, Hoogmoed MS, Lescure J (1977) Etude taxinomique de *Rhinoclemmys punctularia punctularia* (Daudin) (Testudinata. Emydidae). Zool Med Leiden 52:63–80

Fritz U, Havas P (2007) Checklist of Chelonians of the world. Vertebr Zool 57:149–368

Fritz U, Barata M, Busack SD, Fritzsch G, Castilho R (2006) Impact of mountain chains, sea straits and peripheral populations ongenetic and taxonomic structure of a freshwater turtle, *Mauremys leprosa*. Zool Scri 35:97–108

Fritz U, Guicking D, Kami H, Arakelyan M, Auer M, Ayaz D, Ayres Fernández C, Bakiev AG, Celani A, Dzŭkič G, Fahd S, Havas P, Joger U, Khabibullin VF, Mazanaeva LF, Sĭroky P, Tripepi S, Valdeón Vélez A, Velo Antón G, Wink M (2007) Mitochondrial phylogeography of European pond turtles (*Emys orbicularis, Emys trinacris*) – an update. Amphibia-Reptilia 28:418–426

Fritz U, Ayaz D, Buschbom J, Kami HG, Mazanaeva LF, Aloufi AA, Auer M, Rifai L, Silic T, Hundsdörfer AK (2008) Go east: phylogeographies of *Mauremys caspica* and *M. rivulata* – discordance of morphology, mitochondrial and nuclear genomic markers and rare hybridization. J Evol Biol 21:527–540

Fritz U, Stuckas H, Vargas-Ramírez M, Hundsdörfer AK, Maran J, Päckert M (2012) Molecular phylogeny of Central and South American slider turtles: implications for biogeography and systematics (Testudines: Emydidae: *Trachemys*). J Zool Syst Evol Res 50:125–136

Gaffney ES (1975) Phylogeny of the chelydrid turtle: a study of shared characters in the skull. Field Geol 33:157–178

Gaffney ES, Bartholomai A (1979) Fossil trionychids of Australia. J Paleontol 53:1354–1360

Gaffney ES, Meylan PA (1988) A phylogeny of turtles. In: Benton ML (ed) The phylogeny and classification of the tetrapods, vol 1. Clarendon Press, Oxford

Gardner JD, Russell AP, Brinkman DB (1995) Systematics and taxonomy of soft-shelled turtles (Family Trionychidae) from the Judith River Group (mid-Campanian) of North America. Can J Earth Sci 32:631–643

Garman S (1891) On a tortoise found in Florida and Cuba, *Cinosternum baurii*. Bull Essex Inst 23:141–144
Geoffroy Saint-Hilaire EF (1809) Memoire sur les tortues molles. Nouv Bull Soc Philom Paris 22:363–367
Gilmore CW (1922) A new fossil turtle, *Kinosternon arizonense,* from Arizona. Proc U S Natl 62:1–8
Gray JE (1825) A synopsis of the genera of Reptiles and Amphibia, with a description of some new species. Ann Philos 10:193–217
Gray JE (1831a) Synopsis Reptilium or short descriptions of the species of reptiles. Part 1. Cataphracta. Tortoises, crocodiles, and enaliosaurians. Treuttel. Wurtz and Co, London
Gray JE (1831b) A specimen of a tortoise regarded as the type of a new genus in the family Emydidae. Proc Zool Soc Lond 1831:106–107
Gray JE (1834) Characters of two new genera of reptiles (*Geoemyda* and *Gehyra*). Proc Zool Soc Lond 1834:99–100
Gray JE (1844) Catalogue of the tortoises, crocodiles, and amphisbaenians in the Collection of the British Museum. Brit Mus Nat Hist, London 1:1–80
Gray JE (1854) Descriptions of a new genus and some new species of tortoises. Proc Zool Soc Lond 1854:133–135
Gray JE (1856a) On some new species of freshwater tortoises from North America, Ceylon and Australia, in the collection of the British Museum. Proc Zool Soc Lond 1956:197–202
Gray JE (1856b) Catalogue of shield reptiles in the collection of the British Museum. Part I. Testudinata (tortoises). Taylor & Francis, London
Gray JE (1860) Description of a new species of *Geoclemmys* from Ecuador. Proc Zool Soc Lond 1860:231–232
Gray JE (1861) On a new species of water-tortoise *(Geoclemmys melanosterna)* from Darien. Proc Zool Soc Lond 186:204–205
Gray JE (1863a) Notes on American Emydidae, and Professor Aggassiz's observations on my catalogue of them. Ann Mag Nat Hist Lond 3:176–183
Gray JE (1863b) Observations on the box tortoises, with the descriptions of three new Asiatic species. Proc Zool Soc Lond 1863:173–177
Gray JE (1864) Revision of the species of Trionychidae found in Asia and Africa, with the descriptions of some new species. Proc Zool Soc Lond 1864:76–98
Gray JE (1869a) Description of *Mauremys laniaria,* a new freshwater tortoise. Proc Zool Soc Lond 1869:499–500
Gray JE (1869b) Notes on the families and genera of tortoises (Testudinata), and on the characters afforded by the study of their skulls. Proc Zool Soc Lond 1869:165–225
Gray JE (1870) Supplement to the catalogue of shield reptiles in the collection of the British Museum. Part I. Testudinata (tortoises). Taylor & Francis, London
Gray JE (1872) Notes on the mud-tortoises of India (Trionyx, Geoffroy). Ann Mag Nat Hist Lond 10:326–340
Gray JE (1873) On a new freshwater tortoise from Borneo (*Orlitia borneensis*). Ann Mag Nat Hist Lond 11:156–157
Griffith E, Pidgeon E (1830) The class Reptilia arranged by the Baron Cuvier, with specific description. In: Griffith E (ed) The animal kingdom arranged in conformity with its organization, by the Baron Cuvier, with additional descriptions of all species Hitherto Named, and of many not before noticed, vol 9, Reptilia. Whitttaker, Treacher, and Co, London
Groombridge B (1982) The IUCN Amphibia-Reptilia Red Data Book. Testudines. Crocodylia. Rhynchocephalia Part 1. IUCN, Gland
Hartweg N (1934) Description of a new kinosternid from Yucatan. Occ Pap Mus Zool Univ Michigan 277:1–2
Hartweg N (1939) A new American *Pseudemys*. Occ Pap Mus Zool Univ Michigan 397:1–4
Hay OP (1899) Description of two new species of tortoises from the tertiary of the United States. Proc U S Nat Mus 22:21–24
Hay OP (1904) On the existing genera of the Trionychidae. Proc Am Phil Soc 42:268–274

Hay OP (1906) Description of two new genera (*Echmatemys* and *Xenochelys*) and two new species (*Xenochelys formosa* and *Terrapene putnami*) of fossil turtles. Bull Am Mus Nat Hist 22:27–31

Hay OP (1908) The fossil turtles of North America. Carnegie Inst Wash Publ 75:1–568

Head JJ, Raza SM, Gingerich PD (1999) *Drazinderetes tethyensis*: a new large trionychid (Reptilia: Testudines) from the marine Eocene Drazinda Formation of the Sulaiman Range, Punjab (Pakistan). Contrib Mus Paleont Univ Michigan 30:199–214

Head JJ, Aguilera OA, Sánchez Villagra M (2006) Past colonization of South America by trionychid turtles: fossil evidence from the Neogene of Margarita Island, Venezuela. J Herpetol 40:378–381

Henderson RW, Hoevers LG (1975) A checklist and key to the amphibians and reptiles of Belize, Central America. Milwaukee Publ Mus Contrib Biol Geol 5:1–61

Hervet S (2003) Deux nouvelles tortues de l'Éocène inferieur de Saint-Papoul (Aude, France). C R Palévol 2:617–624

Hervet S (2004a) Systématique du Groupe "*Palaeochelys* sensu lato -*Mauremys*" (Chelonii, Testudinoidea) du Tertiaire d'Ëurope occidentale. Ann Paléontol 90:13–78

Hervet S (2004b) A new genus of "Ptychogasteridae" (Chelonii, Testudinoidea) from Geiseltal (Lutetian of Germany). C R Palevol 3:125–132

Hirayama R (1984) Cladistic analysis of batagurine turtles (Batagurinae: Emydinae: Testudinoidea): A preliminary result. Stud Geol Salmant, vol Esp 1 Stud Palaeochel 1:141–157

Holbrook JE (1836) North American herpetology; or a description of the reptiles inhabiting the United States, vol 1. Dobson J and Son, Philadelphia

Honda M, Yasukawa Y, Hirayama R, Ota H (2002a) Phylogenetic relationships of the Asian box turtles of the genus *Cuora* sensu lato (Reptilia: Bataguridae) inferred from mitochondrial DNA sequences. Zool Sci 17:1305–1312

Honda M, Yasukawa Y, Ota H (2002b) Phylogeny of the Eurasian freshwater turtles of the genus *Mauremys* Gray 1869 (Testudines), with special reference to a close affinity of *Mauremys* japonica with *Chinemys reevesii*. J Zool Syst Evol Res 40:195–200

Hutchison LE (1991) Early Kinosteminae (Reptilia:Testudines) and their phylogenetic significance. J Vertebr Paleontol 11:145–167

Hutchison JH (2013) New turtles from the Paleogene of North America. In: Brinkman DB, Holroyd PA, Gardner JD (eds) Morphology and evolution of turtles, vol 26, pp 477–497. Proceedings of Gaffney Turtle Symposium (2009) in Honor of Eugene Gaffney, Vertebrate Paleobiology and Paleoanthropology, Springer, Dordrecht

Hutchison JH, Bramble DM (1981) Homology of the plastral scales of the Kinosternidae and related tunks. Herpetologica 37:73–85

Hutchison JH, Holroyd PA (2003) Late Cretaceous and early Paleocene turtles of the Denver Basin, Colorado. Rocky Mt Geol 38:121–142

Iverson JB (1979) A taxonomic reappraisal of the yellow mud turtles. *Kinosternon flavescens* (Testudines: Kinosternidae). Copeia 1979:212–225

Iverson JB (1981) Kinosternon dunni. Cat Am Amphibia-Reptilia 278:1–2

Iverson JB (1985) Checklist of the turtles of the world with English common names. SSAR Herp Circ 14:1–14

Iverson JB (1988) Neural bone patterns and the phylogeny of the turtles of the subfamily Kinosterninae. Pub Mus Contrib Biol Geo 75:1–12

Iverson JB (1991) Phylogenetic hypotheses for the evolution of modern kinosternine turtles. Herpetol Monogr 5:1–27

Iverson JB (1992) A revised checklist with distribution maps of the turtles of the world. Privately Printed, Richmond

Iverson JB (1998) Molecules, morphology, and mud turtle phylogenetics. Chelon Conserv Biol 3:113–117

Iverson JB, Carr JL, Castaño Mora OV, Galvis Rizo CA, Rentería Moreno LE, Forero-Medina G (2012) *kinosternon dunni* Schmidt 1947 Dunn's Mud turtle, Cabeza de Trozo. In: Rhodin AGJ, Pritchard PCH, van Dijk PP, Saumure RA, Buhlmann KA, Iverson JB, Mittermeier RA (eds)

Conservation biology of freshwater turtles and tortoises: a compilation project of the IUCN/SSC tortoise and freshwater turtle specialist group, vol 5, Chelonian research monographs. Chelonian Research Foundation, Lunenburg, pp 067.1–067.5

Joyce WG, Lyson TR (2011) New material of *Gilmoremys lancensis* nov comb (Testudines: Trionychidae) from the Hell Creek Formation and the diagnosis of plastomenid turtles. J Paleontol 85:442–459

Joyce WG, Revan A, Lyson TR, Danilov IG (2009) Two new plastomenine softshell turtles from the Paleocene of Montana. Bull Peabody Mus Nat Hist 50:307–325

Khosatzky LI, Mlynarski M (1966) Fossil tortoises of the genus *Geoemyda* Gray, 1834 (s. lat) of Europe. Acta Zool Cracoviensia 11:397–421

Kordikova EG (1994a) Review of fossil trionychid localities in the Soviet Union. Courier Forschungs-Institut Senckenberg 173:341–358

Kordikova EG (1994b) About systematics of fossil trionychids in Kazakhstan. Selevinia 2:3–8

Krenz JG, Naylor GJP, Shaffer HB, Janzen FJ (2005) Molecular phylogenetics and evolution of turtles. Mol Phylogenet Evol 37:178–191

Lacépède BGE (1788) Histoire naturalle des quadrupedes ovipaires et des serpens, vol 1. lmprimerie du Roi, Paris

Lamb T, Lydeard C (1994) A molecular phylogeny of the Gopher tortoises, with comments on familial relationships within the Testudinoidea, with comments on familial relationships within the Testudinoidea. Mol Phylgenet Evol 3:283–291

Lapparent de Broin F (2001) The European turtle fauna from the Triassic to the present. Dumerilia 4:155–217

Le M, McCord W (2008) Phylogenetic relationships and biogeographical history of the genus *Rhinoclemmys* Fitzinger, 1835 and the monophyly of the turtle family Geoemydidae (Testudines: Testudinoidea). Zool J Linn Soc 153:751–767

LeConte J (1854) Description of four new species of *Kinosternum*. Proc Acad Nat Sci Phila 7:180–190

Legler JM (1960) A new subspecies of slider turtle (*Pseudemys scripta*) from Coahuila, Mexico. Univ Kansas Publ Mus Nat Hist 13:73–84

Legler JM (1965) A new species of turtle genus *Kinosternon* from Central America. Univ Kansas Publ Mus Nat Hist 15:615–625

Legler JM (1990) The genus *Pseudemys* in Mesoamerica: taxonomy, distribution and origins. In: Gibbons JW (ed) Life history and ecology of the slider turtle. Smithsonian Institution Press, Washington, DC

Legler JM, Webb RG (1970) A new slider turtle (*Pseudemys scripta*) from Sonora, Mexico. Herpetologica 26:157–168

Lindholm WA (1929) Revidiertes Verzeichnis der Gattungen der rezenten Schildkröten nebst Notizen zur Nomenklatur einiger Arten. Zool Anz 81:275–295

Lindholm WA (1931) Uber eine angeblichte *Testudo*-Art aus Sudchina. Zool Anz 97:27–30

Linnaeus C (1758) Systema Naturae per regna tria naturae, secundum classes, ordines, genera, species, cum characteribus, differentiis, synonymis, locis. Editio decima, reformata. Tomus I. Laurentii Salvii, Holmiae

Linnaeus C (1766) Systema Naturae per Regna tria Naturae secundum Classes. Ordines. Genera. Species. cum Charcteribus. Differentiis. Synonymis. Locis. 12th ed. Halae Magdeborgicae

Loomis FB (1904) Two new river reptiles from the Titanothere beds. Am J Sci 18:427–432

Matthew WD (1924) Third contribution to the Snake Creek fauna. Bull Am Mus Nat Hist 50:59–210

McCord WP, Iverson JB, Spinks PQ, Shaffer HB (2000) A new genus of geoemydid turtle from Asia. Hamadryad 25:20–24

McCord WP, Joseph-Ouni M, Hagen C, Blanck T (2010) Three new subspecies of *Trachemys venusta* (Testudines: Emydidae) from Honduras Northern Yucatan, and Pacific Coastal Panama. Reptilia 71:39–49

McDowell SB (1964) Partition of the genus *Clemmys* and related problems in the taxonomy of aquatic Testudinidae. Proc Zool Soc Lond 143:239–279

Medem F (1961) Contribuciones al conocimiento sobre la morfología, ecología y distribución geográfica de la tortuga *Kinosternon dunni* K.P. Schmidt. Novedades Colombianas 1:446–476

Medem F (1962) La distribución geográfica y ecología de los Crocodylia y Testudinata en el Departamento del Choco. Rev Acad Colomb Cien Ex Fis Nat 11:279–303

Medem F (1977) Contribucion al conocimiento sobre la taxonomía, distribución geográfica y ecología de la tortuga "Bache" (*Chelydra serpentina acutirostris*). Caldasia 12:41–101

Merrem B (1820) Versuch eines Systems der Amphibien, Tentamen Systematic Amphiborum. Krieger, Marburg

Mertens R (1954) Zur Kenntnis der Schildkrotenfauna Venezuelas. Senck Biol 35:3–7

Meylan PA (1987) The phylogenetic relationships of soft-shelled turtles (Family Trionychidae). Bull Am Nat Hist 186:1–101

Meylan PA, Gaffney ES (1989) The skeletal morphology of the Cretaceous Cryptodiran turtle, *Adocus*, and the relationships of the Trionychoidea. Am Mus Novit 2941:1–60

Meylan PA, Weig BS, Wood RC (1990) Fossil soft-shelled turtles (Family Trionychidae) of the Lake Turkana Basin, Africa. Copeia 1990:508–528

Moll EO, Legler JM (1971) The life history of a Neotropical slider turtle, *Pseudemys scripta* (Schoepff), in Panama. Bull Los Angel Co Mus Nat Hist Sci 11:1–102

Paolillo OA (1985) Description of a new subspecies of the turtle *Rhinoclemmys punctularia* (Daudin) (Testudines Emydidae) from southern Venezuela. Amphibia-Reptilia 6:293–305

Parhan JF, Feldman CR, Boore J (2006) The complete mitochondrial genome of the enigmatic bigheaded turtle (*Platysternon*): description of unusual genomic features and the reconciliation of phylogenetic hypotheses based on mitochondrial and nuclear DNA. BMC Evol Biol. doi:10.1186/14712148611

Peters WKH (1854) Obersicht der auf seiner Reise nach Mossambique beobachteten Schildkroten. Monatsber Acad Wiss Berl 1854:215–216

Peters WKH (1862) Über einen neuen *Phylodactylus* aus Guayaquil. Monatsber Akad Wiss Berl 1862:626–627

Peters WKH (1868) Eine Miltheilung über eine neue Nagergattung. *Chiropodomys pencillatus,* so wie tiber einige neue oder weniger bekannte Amphibien und Fische. Amphibien. Monatsber Abd Wiss Berl 1868:448–453

Phillipps CA, Dimmick WW, Carr J (1996) Conservation genetic of the common snaping turtle (*Chelydra serpentina*). Conserv Biol 10:397–405

Pidoplichko IG, Tarashchuk BI (1960) New genus of large-headed (macrocephalous) turtle from the Pontion beds in the environs of Odessa. Zbirn Prats Zool Mus Akad USSR 29:105–110 (In Ukranian)

Pomel A (1847) Note sur les animaux fossiles découverts dans le département de l'Allier. Bull Soc Géol Fr 4:378–385

Praschag P, Schmidt C, Fritzsch G, Müller A, Gemel R, Fritz U (2006) *Geoemyda silvatica*, an enigmatic turtle of the Geoemydidae (Reptilia: Testudines), represents a distinct genus. Org Divers Evol 6:151–162

Pritchard PCH (1964) Turtles of the British Guiana. J Brit Guiana Mus 39:19–45

Pritchard PCH (1967) Turtles of the world. T F H Publications, Jersey City

Pritchard PCH (1979) Encyclopedia of Turtles. T F H Publ Neptune New Jersey

Pritchard PCH (1984) Evolution and zoogeography of South American turtles. Studia Geol Salmant Vol Esp 1 Stud Palaeochel 1:225–233

Pritchard PCH, Trebbau P (1984) The turtles of Venezuela. SSAR Contrib Herpetol Ithaca, New York

Rafinesque CS (1815) Analyse de la Nature ou Tableau de l'Univers et des Corps Organises. Privately printed, Palermo

Rafinesque CS (1832) Description of two new genera of soft shell turtles of North America. Atl J Fr Knowl 1:87–110

Reinach AV (1900) Schildkötenreste im Mainzer Tertiärbecken und in benachbarten, ungefahrgleiterigen Ablagerungen. Abh heraus Senckenberg Natur Gesell Frankfurt 28:1–135

Rhodin AGJ, Van Dijk PP, Iverson JB, Broadley Shaffer J (2010) Turtles of the world, 2010 update: annoted checklist of taxonomy, synonymy, distribution, and conservation status. In: Rhodin AGJ, Pritchard PCH, van Djik PP, Saumure RA, Buhlman KA, Iverson JB, Mittermeier RA (eds) Conservation biology of freshwater turtles and tortoises: a compilation project of the

IUCN/SSC tortoises and freshwater turtle specialist group, vol 5, Chelonian research monographs. Chelonian Research Foundation, Lunenburg, pp 000.85–000.164
Ritgen FA (1828) Versuch einer Naturlichen Eintheilung der Amphibien. Nov Acta Physico-Medica Acad Caes Leopold Carol Natur Curio 14:246–284
Rogner M (1996) Schildkröten 2. Hürtgenwald. Heidi, Rogner, Verlag
Sánchez Villagra MR, Asher RJ, Rincon AD, Carlini AA, Meylan PA, Purdy RA (2004) New faunal reports for the Cerro La Cruz locality (Lower Miocene), north western Venezuela. In: Sánchez Villagra MR, Clark JA (eds) Fossil of the Miocene Castillo formation, Venezuela: contributions on neotropical palaeontology, vol 71, Special papers in palaeontology. Palaeontological Association, London, pp 105–112
Savage JM (1966) The origins and history of the Central American herpetofauna. Copeia 1966:719–766
Schleich HH (1994) Neue Reptilienfunde aus dem Tertiär Deutschlands. 13. Schildkröten und Krokodilreste aus der eozänen Braunkohle des Untertagebaues Stolzenbach bei Borken (Hessen) (Reptilia: Crocodylia, Testudines). Courier Forsch Senckenberg 173:79–101
Schmidt KP (1928) Amphibians and land reptiles of Puerto Rico with a list of those reported from the Virgin Islands. In: Scientific survey of Porto Rico and theVirgin Islands. New York Acad Sci 10:1–160
Schmidt KP (1945) A new turtle from the Paleocene of Colorado. Field Geol 10:1–4
Schmidt KP (1947) A new kinosternid turtle from Colombia. Field Zool 31:109–112
Schoepff JD (1792–1801) Historia Testudinum iconibus Illustrata. Palmii Erlangae
Schweigger AF (1812) Prodomus Monographiae Chelooniorum. Königsberg Arch Naturwiss Math 1:171–368
Seidel ME (1989) *Trachemys dorbigni* (Duméril and Bribron) d´Orbigny slider. Cat Am Amphib Reptil 486:1–3
Seidel ME (1990) Trachemys dorbigni. Cat Am Amphib Reptil 486:1–3
Seidel ME (2002) Taxonomic observations on extant species and subspecies of slider turtles, genus *Trachemys*. J Herpetol 36:285–292
Seidel ME, Jackson DR (1990) Evolution and fossil relationship of slider turtles. In: Gibbons JW (ed) Life history and ecology of the slider turtle. Smithsonian Institution Press, Washington, DC
Seidel ME, Iverson JB, Adkins MD (1986) Biochemical comparisons and phylogenetic relationships in the family Kinosternidae (Testudines). Copeia 1986:285–294
Shaffer BH, Meylan P, McKnight M (1997) Tests of turtle phylogeny: molecular, morphological, and paleontological approaches. Syst Biol 46:235–268
Simpson GG (1943) Turtles and the origin of the fauna of Latin America. Am J Sci 241:413–429
Smith MA (1931) The fauna of British India, including Ceylan and Burma. Reptilia and Amphibia 1, Loricata, Testudines. Taylor & Francis, London
Smith HM, Smith RB (1979) Synopsis of the herpetofauna of Mexico, vol VI. Guide to Mexican turtles. Johnson J (ed) North Bennington, Vermont 1044p.
Spinks PQ, Broadley F, Iverson JB, McCord WP (2004) Phylogenetic hypotheses for the turtle family Geoemydidae. Mol Phylogenet Evol 32:164–182
Spix JB von (1824) Animalia nova. species novae Testudinum et Ranarum quas in itinere per Brasiliam ann is 1817–1820 collegit et descripsit. Hubschmann, München
Stejneger LH (1902) Some generic names of turtles. Proc Biol Soc Wash 15:235–238
Stejneger LH (1925) New species and subspecies of American turtles. J Wash Acad Sci 15:462–463
Stuart LC (1963) A checklist of the herpetofauna of Guatemala. Misc Publ Mus Zool Univ Michigan 122:1–150
Theobald W Jr (1868) Catalogue of Reptiles in the Museum of the Asiatic Society of Bengal. J Asiatic Soc Extra Number, 1–88
Vanzolini PE (1995) A new species of turtle genus *Trachemys* from the State of Maranhao, Brazil (Testudines, Emydidae). Rev Brasil Biol 55:111–125
Vanzolini PE, Heyer WR (1985) The American Herpetofauna and the interchange. In: Stheli FG, Webb SD (eds) The great American biotic interchange. Plenum Press, New York/London

von Meyer H (1847) *Palaeochelys bussenensis* im älteren Süsswasserkalk. Jahre Vereins Vaterlán Natur Württenberg 3:167–168

Wagler J (1830) Natürliches System der Amphibien mit Vorangehander Classification der Säugthiere und Vögel. 1. G. Cotteschen Buchhandlung, München

Whetstone KN (1978) A new genus of cryptodiran turtles (Testudinoidea, Chelydridae) from the Upper Cretaceous Hell Creek Formation on Montana. Univ Kansas Sci Bull 51:539–563

White AW (1998) A new Eocene soft-shelled turtle (Tryonichidae) from Murgon, South-eastern Queensland. Mem Assoc Austral Palaeont 25:37–44

Williams EE (1952) A staurotypinae skull from the Oligocene of South Dakota. Breviora 2:1–16

Wood RC (1979) First record of a fossil trionychid skull from Africa. Herpetologica 35:360–364

Wood RC (1997) Turtles. In: Kay RF, Madden RH, Cifelli RL, Flynn JJ (eds) Vertebrate paleontology in the neotropics. The Miocene Fauna of La Venta, Colombia. Smithsonian Institution Press, Washington, DC

Wood RC, Patterson B (1973) A fossil trionychid turtle from South America. Breviora 405:1–10

Woodburne MO (2010) The great American biotic interchange: dispersals, tectonics, climate, sea level and holding pens. J Mamm Evol 17:245–264

Yasukawa Y, Hirayama R, Hikida T (2001) Phylogenetic relationships of geoemydine turtles (Reptilia: Bataguridae). Curr Herpetol 20:105–133

Yeh HK (1963) Fossil turtles of China. Palaeontol Sin 150:1–112

Chapter 3
Tortoises: The Oldest Island Hoppers

Abstract Tortoises are terrestrial turtles grouped in the clade Testudinidae. This clade represents 18 % of the living turtle diversity and is the most widespread group of nonmarine turtles, distributed across all nonpolar continents except Australia. The fossil history of the clade dates back to the Palaeocene of Asia. The South American forms are represented by the genus *Chelonoidis*, which includes 4 extant and 4 extinct named species. The invasion of tortoises in South American is a contentious subject, and two continental sources have been proposed (Central America and Africa).

Keywords Testudinidae • *Chelonoidis* • Oligocene • Neogene • Biogeography

3.1 Testudinidae Batsch 1788

The true tortoises are included in the clade Testudinidae. This group is the second richest clade among living turtles (almost 50 species after Ernst and Barbour 1989; Rhodin et al. 2010). According to Auffenberg (1974), its extant members are located mainly in subtropical to tropical areas, and they are most common in subhumid to arid grassland and savanna habitats in portions of Africa, Aldabra atoll, Galápagos Islands, India, Madagascar, South America, southeast Asia, southern Europe, and southern North America (see distribution maps in Auffenberg 1974 and Ernst and Barbour 1989; Iverson 1992) (Fig. 3.1a). However, the clade had a wider distribution in the past as shown by its fossil record; Testudinidae were found in additional regions of North America and southern Canada, northern Europe, Central Asia, and the West Indies (Auffenberg 1974, Fig. 1). Pritchard and Trebbau (1984) pointed out that Testudinidae have successfully occupied many distinct terrestrial

M.S. de la Fuente et al., *Origin, Evolution and Biogeographic History of South American Turtles*, Springer Earth System Sciences, DOI 10.1007/978-3-319-00518-8_3,

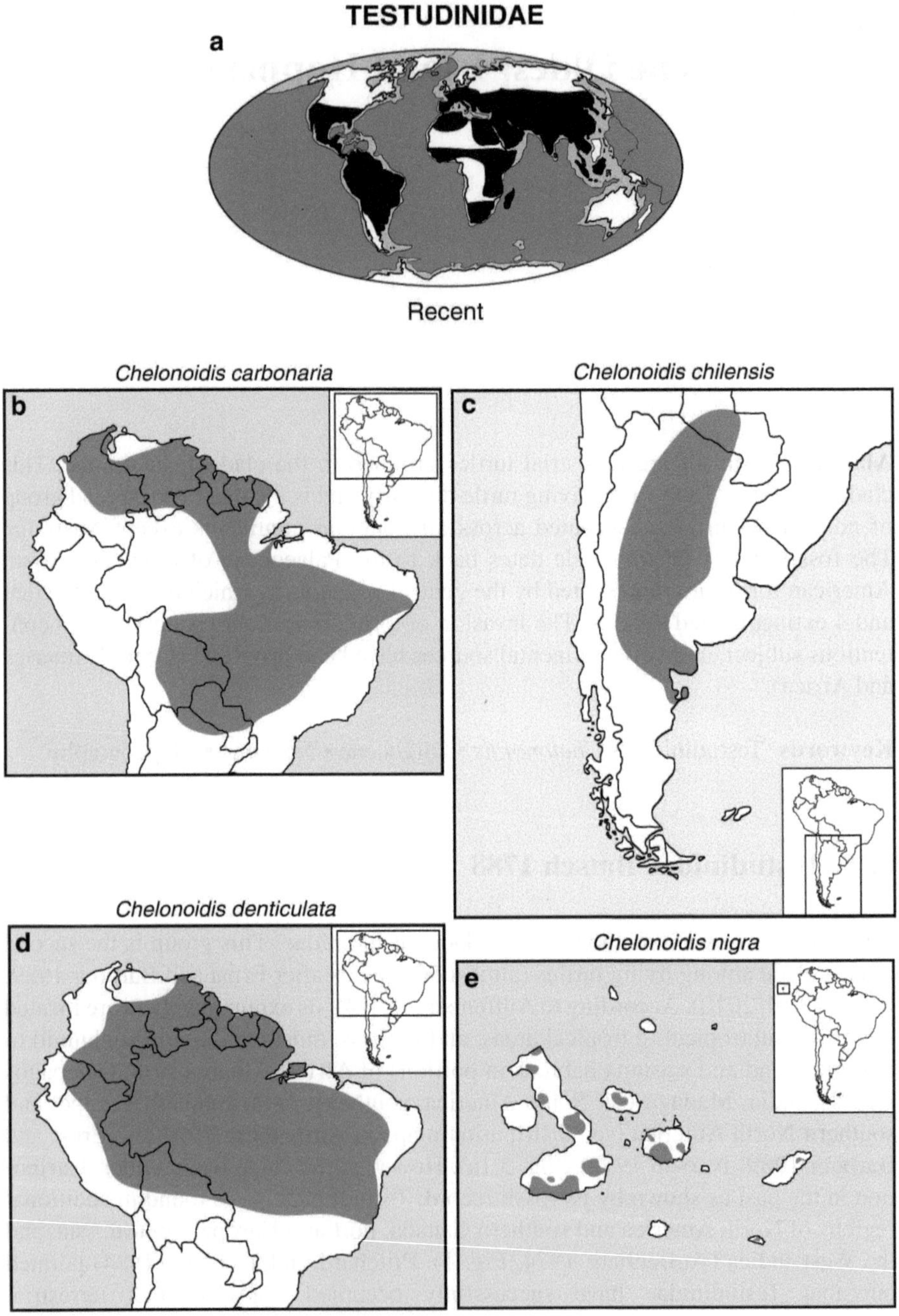

Fig. 3.1 Geographic distribution of Testudinidae in recent times. (**a**) Global distribution of Testudinidae. (**b**) Geographic distribution of *Chelonoidis carbonaria* in South America. (**c**) Geographic distribution of *Chelonoidis chilensis* in South America. (**d**) Geographic distribution of *Chelonoidis denticulata* in South America. (**e**) Geographic distribution of *Chelonoidis nigra* in South America

ecosystems and have evolved in 11 different morphotypes[1]: (1) generalized xeric tropical forms (e.g., species included in *Astrochelys* and *Geochelone*); (2) generalized xeric temperate forms (e.g., extant and extinct species of *Testudo*); (3) dwarf xeric, subtropical forms (e.g., *Homopus*, *Psammobates*); (4) species with kinetic shells (e.g., *Kinixys*, *Pyxis arachnoides*); (5) specialized rock-dwelling species (e.g., *Malacochersus tornieri*); (6) burrowing species (e.g., *Gopherus*); (7) highly armed species (e.g., *Hesperotestudo*); (8) thick-shelled "giant" tortoises (e.g., large *Chelonoidis* extinct species); (9) thin-shelled "giant" tortoises (e.g., *Chelonoidis nigra*); (10) "supergiant" continental tortoises (e.g., *Geochelone bolivari*); and (11) humid forest species (e.g., *Chelonoidis denticulata*, *Manouria emys*).

The oldest extinct Testudinidae have been discovered from the Early Palaeocene of Asia (Claude and Tong 2004; Hutchison 1998; Joyce et al. 2004), and the Lindholmemydidae, the closest relatives to Testudinoidea, have been found in the Late Cretaceous of Asia (Sukhanov 2000, and references therein). From Asia, the tortoises dispersed into other continents as documented by their fossil record in the Early Eocene of Europe and North America and in the Late Eocene of Africa (Claude and Tong 2004; Crumly 1984a; Holroyd and Parham 2003; Hutchison 1996, 1998).

Although the monophyly of Testudinidae has never been disputed and was supported by morphological and molecular analyses (Crumly 1984a; Gaffney and Meylan 1988; Gerlach 2001; Krenz et al. 2005; Near et al. 2005; Spinks et al. 2004; van der Kuyl et al. 2002; Le et al. 2006), several papers (Crumly 1984a; Gaffney and Meylan 1988; Caccone et al. 1999a; Meylan and Sterrer 2000; Gerlach 2001; Palkovacs et al. 2002) have resulted in conflicting hypotheses concerning the mono-, para-, or polyphyly of some particular genera (e.g., *Geochelone*). Most molecular analyses have had a restricted taxonomic sampling (Lamb and Lydeard 1994; Caccone et al. 1999a, b; Austin and Arnold 2001; Austin et al. 2003; Iverson et al. 2001; Palkovacs et al. 2002; Fritz et al. 2005; Parham et al. 2006), because they were conducted to resolve specific relationships among Testudinidae. Recently, broader sampling by Le et al. (2006) and Fritz and Bininda-Emonds (2007) resulted in more robust and resolved phylogenies.

Genera included: *Agrionemys* Khosatzky and Mlynarsky 1966, *Astrochelys* Gray 1873, *Cheirogaster* Bergounioux 1935, *Chelonoidis* Fitzinger 1835, *Chersina* Gray 1831, *Cylindraspis* Fitzinger 1835, *Dipsochelys* Bour 1982, *Dithyrosternon* Pictet and Humbert 1855–1857, *Ergilemys* Chkhikvadze 1972, *Geochelone* Fitzinger 1835, *Gopherus* Rafinesque 1832, *Hadrianus* Cope 1872b, *Hesperotestudo* Williams 1950, *Homopus* Duméril and Bibron 1834, *Indotestudo* Lindholm 1929, *Impregnochelys* Meylan and Auffenberg 1986, *Kinixys* Bell 1827, *Malacochersus* Lindholm 1929, *Manouria* Gray 1854a, *Palaeotestudo* Lapparent de Broin 2000, *Psammobates* Fitzinger 1835, *Pyxis* Bell 1827, *Stigmochelys* Gray 1873, *Stylemys* Leidy 1851, and *Testudo* Linnaeus 1758.

[1]Courtesy Society for the Study of Amphibians and Reptiles.

South American Genera and Species

3.1.1 *Chelonoidis* Fitzinger 1835

Type species: *Testudo denticulata* Linnaeus 1766
Definition: See Auffenberg (1974).
Species included: *Chelonoidis australis* (Moreno 1889), *Ch. carbonaria* (Spix 1824), *Ch. chilensis* (Gray 1870), *Ch. denticulata* (Linnaeus 1776), *Ch. gallardoi* (Moreno 1889), *Ch. gringorum* (Simpson 1942), *Ch. hesterna* (Auffenberg 1971), and *Ch. nigra* (Quay and Gaimard 1824).

Chelonoidis carbonaria (Spix 1824)

Holotype: According to King and Burke (1989), it was housed originally in Zoologisches Sammlung des Bayerischen Staates (ZSM), Münich, Germany, but it is presently lost.
Diagnosis: See Pritchard and Trebbau (1984).
Distribution: Around the Amazon River Basin from Panama and Colombia to Paraguay, Bolivia, northeastern Argentina, and Brazil (Pritchard and Trebbau 1984, Iverson 1992) (Fig. 3.1b).
Comments: See *Ch. denticulata.*

Chelonoidis chilensis (Gray 1870)

Holotype: 2 syntypes housed in the British Museum of Natural History, London, United Kingdom. BMNH 70.12.18.2.
Definition: See Ernst and Barbour (1989).
Distribution: The Espinal and Monte Regions of Argentina and the Paraguayan Chaco (Fernández 1988) (Fig. 3.1c).
Comments: The number of species in the *Chelonoidis chilensis* complex is debated. Traditionally, the small-sized Argentinean and Paraguayan tortoise was referred to only one species, *Chelonoidis chilensis* (Gray 1873). More recently, Freiberg (1973) described two additional species: *Ch. petersi* from the northern range (Chaco) of the *Ch. chilensis* complex and *Ch. donosobarrosi* from the southernmost part of the range (northern Patagonia). Freiberg's (1973) species were defined by differences in size (*Ch. petersi* is small while *Ch. donosobarrosi* is medium-sized) and minor coloration and additional shell differences (i.e., *Ch. petersi* has yellow central carapace and smooth plastral scutes, while *Ch. donosobarrosi* has smooth carapacial scutes with black centers and growth rings on the plastral scutes). Both species plus *Ch. chilensis* were recognized by Cei (1986), while Buskirk (1993) synonimized *Ch. donosobarrosi* with *Ch. chilensis.* Fernández (1988) and Cabrera (1998) recognized only two species: *Ch. petersi* (restricted to northern populations) and *Ch. chilensis* (restricted to southern ones). These authors pointed out that *Ch. donosobarrosi* is a junior synonymous of *Ch. chilensis.* Although Richard (1999) recognized two species, the designation is different from Fernández (1988) and Cabrera (1998). Richard

(1999) recognized *Ch. chilensis* (northern populations), including *Ch. petersi* as a synonym, and *Ch. donosobarrosi* (southern populations). In contrast with these taxonomic decisions, Ernst and Barbour (1989) and Ernst et al. (2000) identified only one species: *Ch. chilensis*, and they have interpreted the diagnostic characters of these three species as a result of north–south clinal or sexually dimorphic trait variation. Using sequence data of the mitochondrial cytochrome b and the length polymorphisms of 10 microsatellite loci, Frits et al. (2012) concluded that only one species (*Chelonoidis chilensis*) is valid among the three nominal species. Accordingly, *Ch. donosobarrosi* (Freiberg 1973) and *Ch. petersi* (Freiberg 1973) would be junior synonyms of *Ch. chilensis* (Gray 1870). The fossil record of *Ch. chilensis* is restricted to outcrops of an unnamed litostratigraphic unit assigned by its mammal faunal assemblage to the Montehermosan SALMA (SALMA: South American Land Mammal Age, Early–Middle Pliocene) of Sierra Chica, Córdoba Province (de la Fuente and Cabrera 1988), and of Carcarañá Formation (Late Pleistocene) near Carcarañá, Santa Fé Province, Argentina (de la Fuente 1997b, 1999).

Chelonoidis denticulata (Linnaeus 1776)

Holotype: Natuthistoriska Riksmusset, Stockolm, Sweden. NRM De Geer collection 21.

Diagnosis: See Pritchard and Trebbau (1984).

Distribution: Northern South America, east of Andes south to southern Brazil (Pritchard and Trebbau 1984; Iverson 1992) (Fig. 3.1d).

Comments: Although Thomas Bell in 1825 (fide Williams 1960) beautifully illustrated and clearly differentiated *Chelonoidis carbonaria* and *Ch. denticulata*, these species were confounded and synonymized for 125 years until Williams (1960) clarified the distinctions between them. This author recognized differences in scale coloration on the carapace, in the morphology of prefrontal and frontal scales, in the concentric growing degree in carapace shields, characteristics of some plastral scales, and the shape and contacts of inguinal scales in adult and juvenile specimens of both species (Williams 1960, Fig. 1 and table in page 2). More recently, the comparative behavior of *Ch. carbonaria* and *Ch. denticulata* was revised by Castaño Mora and Lugo (1981). Finally, Pritchard and Trebbau (1984) have presented a detailed revision of both species. In this context, these authors described the external morphology and the osteology of the shell and skull, coloration, and extremities of both species. Peter Pritchard and Pedro Trebbau also summarized the available information regarding their size and growth, distribution, geographic variation, habitat, feeding, reproduction, economic importance, and vernacular names. The fossil record of *Ch. denticulata* is reported by Manzano et al. (2009) from the Arroyo Feliciano Formation (Late Pleistocene), which crops out along major rivers and streams in Entre Ríos Province, Argentina (Iriondo et al. 1985; Iriondo 1996). According to Manzano et al. (2009), this record represents the range extension of this species in 800 km south from its extant distribution. This is in accordance to subtropical temperate to warm and humid climate during the last interglacial period that characterized Entre Ríos Province, northeastern Argentina.

Chelonoidis nigra (Quay and Gaimard 1824)

Holotype: Muséum National D'Histoire Naturelle, Paris, France. MNHN 9550.
Definition: See Ernst and Barbour (1989).
Distribution: Galápagos Island, Ecuador (Iverson 1992) (Fig. 3.1e).
Comments: Bour (1984) showed that *Testudo nigra* Quay and Gaimard 1824 is the correct name for the famous tortoises of Galápagos Islands, rather than *Testudo elephantopus* Harlam 1827. Within the Galápagos Archipelago, ten subspecies of *Ch. nigra* have been recognized, but only seven survive presently (Pritchard 1996). Most of them have occurred on separate islands (Española, Fernandina, San Cristóbal, San Salvador, Santa Cruz, Santa María, Pinta, and Pinzón), and the remaining five separate populations are living on the slopes of different volcanoes on Isabella Island. In a more recent taxonomic list, Rhodin et al. (2010) elevated the *nigra* subspecies to the species level [*Ch. nigra* (Quay and Gaimard 1824), *Ch. abingdonii* (Günther 1877), *Ch. becki* (Rothschild 1901), *Ch. chatanensis* (Van Denburgh 1907), *Ch. darwini* (Van Denburgh 1907), *Ch. duncanensis* (Garman 1917), *Ch. hoodensis* (Van Denburgh 1907), *Ch. phantastica* (Van Denburgh 1907), *Ch. porteri* (Rothschild 1903), *Ch. vicina* (Günther 1875)] including these species in the *Chelonoidis nigra* complex. Although this taxonomic decision may be acceptable, we concur with King and Burke (1989) and the Galápagos tortoises are treated in this monograph as a nominal species (*Ch. nigra*) with ten subspecies, because the Galápagos Islands populations are more closely related to each other than they are to mainland species. This interpretation is based on morphologic (Crumly 1982; Fritts 1983), electrophoretic (Marlow and Patton 1981), and molecular (Caccone et al. 1999b) studies.

Despite their close relationships, the tortoises from the Galápagos Islands show great shell diversity (Fritts 1983). Two basic carapace morphologies are recognized in *Ch. nigra*: domed and saddleback. Although it was thought that saddleback carapaces derived from domed ones, Fritts (1983) has suggested that saddleback morphology evolved more than once. According to Crumly (1984b), the shell plasticity that is shown in different subspecies of *Ch. nigra* is not recognized in skull morphology.

The origin and phylogenetic relationships of Galápagos tortoises was addressed by Caccone et al. (1999b). In accordance with their phylogenetic analyses, which are based on mitochondrial DNA sequences from *Ch. nigra* subspecies and *Chelonoidis* species of mainland South America (*Ch. carbonaria*, *Ch. denticulata*, and *Ch. chilensis*), the closest relative to *Ch. nigra* is the small Chaco tortoise (*Chelonoidis chilensis*). According to Caccone et al. (1999b), the timing of the split between *Ch. nigra* and *Ch. chilensis* occurred before the origin of the Galápagos Islands (between six and two million years ago).

Chelonoidis australis (Moreno 1889)

Holotype: Museo de La Plata, La Plata, Argentina. MLP 26-400, carapace.
Diagnosis: See de la Fuente (1997a).
Locality, horizon, and age: Atlantic Ocean coast between Monte Hermoso and Punta Alta, Buenos Aires Province, Argentina. Monte Hermoso Formation (SALMA Montehermosan, Early–Middle Pliocene) (Fig. 3.2a, b).
Comments: When Rovereto (1914) named two new testudinids from the Cenozoic of Argentina (*Testudo gallardoi* and *T. praestans*), he did not mention Moreno (1889) who had previously erected two species (*Testudo australis* and *T. formosa*) from the Monte Hermoso Formation. The specimen that Moreno used to name *Testudo australis* was found in the Department of Vertebrate Paleontology at the Museo de La Plata, and it was described by de la Fuente (1988, 1997a). Unfortunately, the specimen named *Testudo formosa* by Pascasio Moreno was not present in the collections of the Museo de La Plata. Because the specimen could not be examined and based on the brief characterization of this species by Moreno (1889), de la Fuente (1997a) referred *Testudo formosa* as a nomen nudum as it was previously suggested by Williams (1950) and Auffenberg (1974). *Testudo australis* was also referred as a nomen nudum by Williams (1950) and included in a list of nomina nuda and unidentifiable material by Auffenberg (1974). However, as pointed out by de la Fuente (1988, 1997a), Williams (1950) did not see Moreno's materials and Auffenberg (1974) did not comment on his decision. According to de la Fuente (1988, 1997a), *Testudo australis* Moreno 1889 is not a nomen nudum because this taxon name is available following the International Code of Zoological Nomenclature and it precedes by 25 years the description of *Testudo praestans*. Concluding, de la Fuente (1988, 1997a) suggested that *T. praestans* should be included as a synonym of *T. australis* and that both should be assigned to the genus *Chelonoidis* sensu Bour (1980).

Chelonoidis hesterna Auffenberg 1971

Holotype: University of California Museum of Paleontology, California, USA. UCMP 40200, a complete shell of an adult male tortoise, crushed dorsally, which contains a badly crushed skull and lower jaw, cervical vertebrae, pelvic girdle, femur, tibia, fibula, most of foot, humerus, and ulna (Auffenberg 1971).
Diagnosis: See Auffenberg (1971).
Locality, horizon, and age: San Nicolás locality, 3 km northeast of Villavieja, Huila Colombia, South America. Honda Group, Cerbatana gravel sandclays, La Venta fauna, Late Miocene (Auffenberg 1971) (Fig. 3.2a, c).
Comments: *Chelonoidis hesterna* was named by Auffenberg (1971) based on the holotype (UCMP 40200) and a referred specimen represented by a partial skull (UCMP 3890) belonging to another specimen that is larger than the holotype. According to Auffenberg (1971), the taxa included in the *carbonaria* group (Williams 1960) are *Ch. carbonaria*, *Ch. denticulata*, and *Ch. hesterna*. The close relationship between *Ch. carbonaria* and *Ch. denticulata* is supported by molecular phylogenies (Le et al. 2006). Likewise, the relationships among the

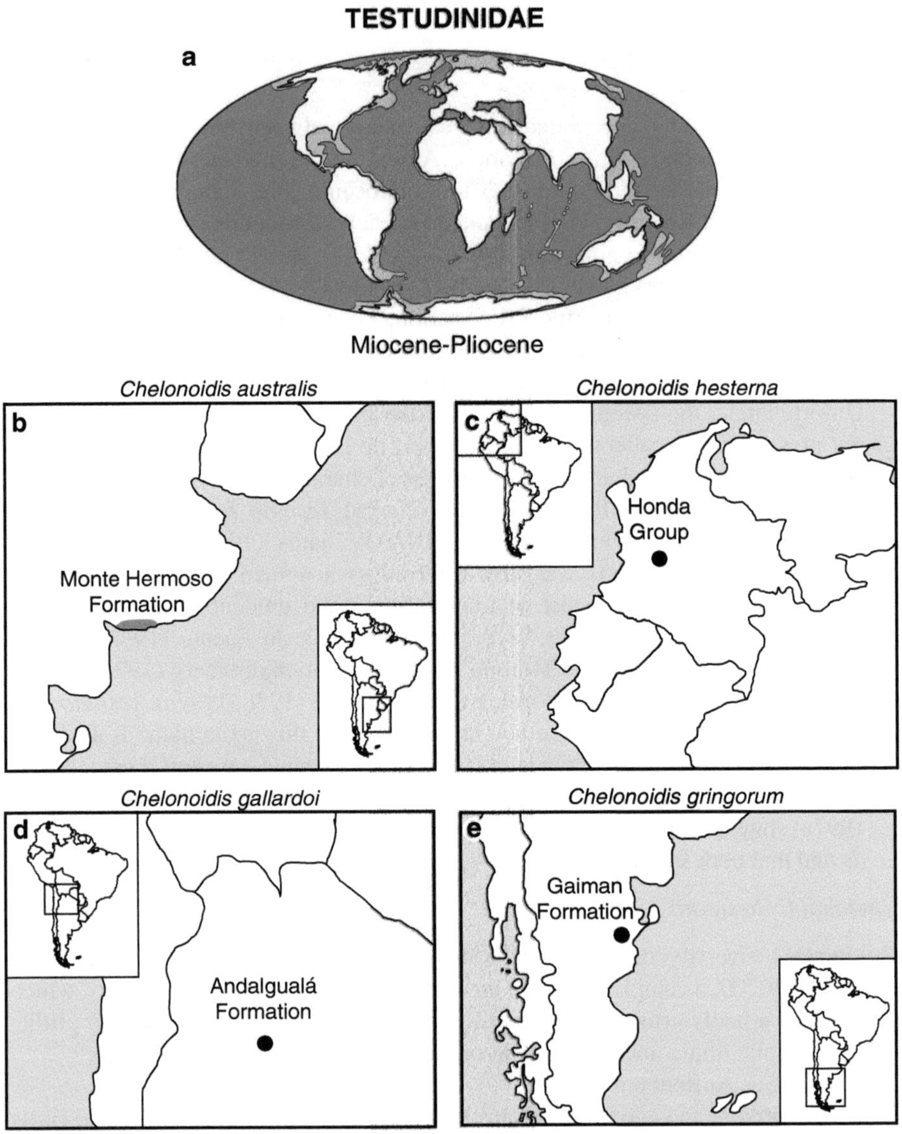

Fig. 3.2 South American record of Testudinidae in the Miocene–Pliocene. (**a**) Paleogeographic reconstruction of continents during the Miocene–Pliocene. (**b**) Geographic and stratigraphic location of *Chelonoidis australis*. (**c**) Geographic and stratigraphic location of *Chelonoidis hesterna*. (**d**) Geographic and stratigraphic location of *Chelonoidis gallardoi*. (**e**) Geographic and stratigraphic location of *Chelonoidis gringorum*

Fig. 3.3 Life reconstruction of *Chelonoidis gallardoi* from the Miocene of the Andalhualá Formation, Argentina

three species are supported by Zacarias et al. (in press). Auffenberg (1971) suggested that *Ch. hesterna* is a basal species of the *carbonaria* group from which the two extant species are derived as a result of the development of a great savannah in the Amazonian region at the end of the Miocene. In de la Fuente's cladogram (de la Fuente 1994), the *carbonaria* group is shown as a polytomy among *Ch. hesterna*, *Ch. carbonaria*, and *Ch. denticulata*. However, in all the most parsimonious trees recovered by Zacarias et al. (in press), a subgroup consisting in *Ch. hesterna*/*Ch. carbonaria* is nested with *Ch. denticulata* in the *carbonaria* group.

Chelonoidis gallardoi (Rovereto 1914) (Fig. 3.3)

Holotype: Museo Argentino de Ciencias Naturales "Bernardino Rivadavia," Buenos Aires, Argentina. MACN 5206, a carapace and plastron of an adult specimen.

Diagnosis: See de la Fuente (1997a).

Locality, horizon, and age: Andalhualá, Catamarca Province, Argentina. Araucanian sensu strict=Andalhualá Formation (Riggs and Patterson 1939; Marshall and Patterson 1981) (Fig 3.2a, d).

Comments: Rovereto (1914) named two new species: *Testudo gallardoi* from the "Araucanian" (= Huayquerian SALMA, Late Miocene) of Andalhualá (Catamarca Province, Argentina) and *T. preastans* from the Montehermosan SALMA (Early–Middle Pliocene) near Monte Hermoso (Buenos Aires Province, Argentina). Rovereto (1914) briefly described both species, illustrating *T. gallardoi*'s carapace with photographs (plate 8, Fig. 6 and plate 10, Fig. 1) and *T. praestans*' carapace and plastron with a schematic drawing (Figs. 69–70). Auffenberg (1974) suggested that *T. preastans* and *T. gallardoi* are conspecific and referred *T. praestans* as a junior synonym of *T. gallardoi*. However, this taxonomic decision was not well

Fig. 3.4 Life reconstruction of *Chelonoidis gringorum* from the Miocene of the Gaiman Formation, Argentina

supported by Auffenberg (1974). This author also assigned *T. gallardoi* to the genus *Geochelone* Fitzinger *sensu* Williams (in Loveridge and Williams 1957) subgenus *Chelonoidis*. More recently, de la Fuente (1988, 1997b) described and illustrated the type specimens of *T. gallardoi* and *T. praestans*, recognized *T. gallardoi* as a full species (as *Ch. gallardoi*), and suggested that *T. praestans* Rovereto 1914 be considered a junior synonym of *Ch. australis* (see above).

Chelonoidis gringorum (Simpson 1942) (Fig. 3.4)

Holotype: American Museum of Natural History, New York, U.S.A. AMNH 3366, a partial carapace and a plastron of one specimen.

Diagnosis: See Simpson (1942) and emended diagnosis in de la Fuente (1988, 1994).

Locality, horizon, and age: South side of the Chubut Valley, between Gaiman and Dolavon (Chubut Province, Argentina), upper section of the "Patagonian" (= Gaiman Formation) and upper section of the Sarmiento Formation (see Mendía and Bayarsky 1981) Late Colhuehuapean SALMA (Early Miocene) (Fig. 3.2a, e).

Comments: *Testudo gringorum* was named by Simpson (1942) based on the holotype that was discovered during field trips developed by the Scarritt Expedition to Patagonia in 1933 from the upper section of the "Patagonian" (Late Colhuehuapean SALMA; Early Miocene) outcropping at the southern side of the Chubut River valley. Firstly, this turtle was referred to the genus *Testudo* Linnaeus by Simpson (1942). Later, according to Williams (in Loveridge and Williams 1957), Auffenberg (1971, 1974) included this species in the genus *Geochelone* Fitzinger (subgenus *Chelonoidis*), and finally, de la Fuente (1988, 1997) included it in the genus *Chelonoidis* sensu Bour (1980). New specimens referred to *Chelonoidis gringorum* by de la Fuente (1988, 1994) were recovered by different paleontological crews in Central Patagonia (see de la Fuente 1994). *Chelonoidis gringorum* was considered "ancestral" to *Ch. chilensis* Gray (Auffenberg 1971) or conforming with YPFB PAL 0932 (from the Deseadan SALMA, Early Oligocene of Bolivia) a clade that represents the sister group of extant South American tortoises (de la Fuente 1994) or the sister group of *Ch. chilensis* and *Ch. nigra* complexes (Zacarias et al. in press).

3.2 Discussion

The South American colonization of the true tortoises of the clade Testudinidae should be analyzed considering the cenogenetic events (Reig 1962, 1968) and the isolation and randomness of insular settlements that have shaped the taxonomic composition of South American tetrapod fauna. In this context it must be noted that these tortoises, along with ceboid monkeys, hystricognathan rodents, scincid lizards, and microhylid frogs, invaded South America by accidental dispersal when South America was isolated (see Simpson 1943, 1980; Reig 1962, 1968, 1981; Patterson and Pascual 1972; Laurent 1979; Pascual 1984a, b). It is interesting to note that the estimated arrival of tortoises in South America (Late Eocene–Early Oligocene) was during its isolation, showing the capacity of tortoises for transoceanic migration. According to Pritchard's assumption, the turtle distribution can be explained considering that freshwater turtles require land bridges for dispersal, but tortoises do not. In this sense, Pritchard (1984) pointed out that (1) tortoises have an impermeable integument and (2) although tortoises are bad swimmers their floating capacity is high. These conditions allowed tortoises to reach distant locations by passive drift. Although Pritchard's rules are correct concerning tortoise dispersal, the idea of land bridges required for the dispersal of fresh water turtles has been refuted by recent discoveries (see Sánchez Villagra et al. 2004; Head et al. 2006; Le and McCord 2008; Cadena 2009, Cadena et al. 2012; Frits et al. 2012). On the other hand, the transoceanic island colonization by testudinid tortoises has been documented on several occasions. Among them *Chelonoidis* in Galápagos Archipelago (Caccone et al. 1999b, 2002; Russello et al. 2005), tortoises from mainland Africa to Madagascar (Caccone et al. 1999a), and *Dipsochelys* and *Cylindraspis* among West Indian Ocean island (Williams 1950, 1952; Bour 1984, 1987; Austin and Arnold 2001; Austin et al. 2003).

Although there is a consensus concerning transoceanic drift for dispersal of South American tortoises, the continental source of these turtles remains as a matter of controversy. Several authors have pointed out that tortoises colonized South America from North and Central America (Simpson 1942, 1943; Williams 1950; Auffenberg 1971; Broin 1991; Gerlach 2001). Contrarily, Simpson (1942) and Crumly (1984a) suggested Africa as an alternative origin. Particularly, Crumly (1984a, c), who does not support the monophyly of South American tortoises, suggested multiple invasions from North and Central America, not discounting an African source for some South American lineages. Supporting Hoffstetter and

Lavocat's (1970) and Lavocat's (1977) hypothesis based on ceboid monkeys and hystricognathan rodents, de la Fuente (1988, 1997b) advanced the hypothesis of an African origin for the *Chelonoidis* lineage. More recently, a biogeographical analysis (Le et al. 2006) based on a molecular phylogeny of Testudinidae that included 23 extant species and both mitochondrial and nuclear DNA data rejected the hypothesis of a North American origin of South American turtles and supported Africa as the ancestral continental area for all testudinids except the genera *Manouria* and *Gopherus*. Le et al. (2006) biogeographic study has shown that the monophyletic group of the South American-Galápagos tortoises is nested in a large clade that includes species from Africa, Asia, and the Indian Ocean. This result is not consistent with the previous hypothesis of a North American origin of *Chelonoidis* species (as *Geochelone carbonaria* clade in their paper) based on sister relationships with *Gopherus* (Gerlach 2001). On the contrary, Le et al. (2006) recovered sister group relationships between *Chelonoidis* species and African *Kinyxis* ones.

According to Hoffstetter (1972, 1974, 1975, 1980) and Lavocat (1969), the dispersal of platyrrhine monkeys and hystricognathan rodents as well as tortoises form Africa to South America during the Late Eocene might have been aided by the Equatorial Current (flow between the Congo Delta and Maranhao, Brazil) and the Trade Winds. These westward currents and winds as well as the relatively smaller distance between Africa and South America during the Eocene and the persistence of the oceanic island far from the South American and Africa coasts (Tarling 1980) support this hypothesis.

Simons (1976) has argued that distances between Africa and South America during the Paleogene were greater than those provided in Hoffstetter and Lavocat's aforementioned papers, and the distance between the Americas was not great enough to preclude faunal interchange between them. Although Tarling (1980) suggested that during the Late Eocene–Early Oligocene, the distances between the Mesoamerican emerged land and South America exceeded 1,200 km, the occurrence of podocnemidids from the Late Eocene–Early Oligocene of the Gatuncillo Formation (Cadena et al. 2012) as well as the Palaeocene records from Colombia (Cadena et al. 2010, 2012) show an earlier invasion of South American turtles across the Panamanian Seaway. Tarling (1980) pointed out that the currents of the Panamanian seaway flow into the Mexican Gulf, precluding migration of South American fauna. To the contrary, the oceanic currents that flow to the west along the northern coast of South America might have reached the Mesoamerican volcanic island and might have aided the dispersal of South American fauna by drift (Tarling 1980).

References

Auffenberg W (1971) A new fossil tortoise, with remarks on the origin of South American Testudinines. Copeia 1:106–117

Auffenberg W (1974) Checklist of fossil land tortoises (Testudinidae). Bull Fla State Mus Biol Sci 18:121–246

Austin JJ, Arnold EN (2001) Ancient mitochondrial DNA and morphology elucidate an extinct island radiation of Indian Ocean giant tortoises (*Cylindraspis*). Proc R Soc Biol Sci Ser B 268:2515–2523

Austin JJ, Arnold EN, Bour R (2003) Was there a second adaptive radiation of giant tortoises in the Indian Ocean? Using mitochondrial DNA to investigate speciation and biogeography of *Aldabrachelys* (Reptilia, Testudinidae). Mol Ecol 12:1415–1424

Batsch AJGC (1788) Versuch einer Anleitung zur Kenntniss und Geschichte der Thiere und Mineralien, vol 1. Akad Buchhandlung, Jena

Bell T (1827) On two new genera of land tortoise having a moveable sternum, with remarks on their arrangement and affinities. Zool J Lond 2:392–401

Bergounioux FM (1935) Contribution à l'étude paléontologique dês Chéloniens, Chéloniens fossiles du Bassin d'Aquitaine. Mém Soc Géol Fr N S 25:1–216

Bour R (1980) Essai sur la taxonomie des Testudinidae actuals (Reptilia, Chelonii). Bull Mus Nat Hist Nat Paris 4:541–546

Bour R (1982) Contribution à la connaissance des tortues terrestres des Seychelles: définition du genre endémique et description d'une espèce nouvelle probablement originaire des îles granitiques et au bord de l'extinction. C R Acad Sci Paris 3:117–122

Bour R (1984) Les tortues terrestres géants des îles de l'Océan Indien Occidental: données géographiques, taxinomiques et phylogénetiques. Stud Palaeochel Salmant 1:17–76

Bour R (1987) Tortues et insularité: les tortues des Seychelles. Bull Soc Zool Fr 112:401–418

Buskirk JR (1993) Distribution, status and biology of the tortoise, *Geochelone chilensis*, in Río Negro Province, Argentina. Stud Neotrop Fauna Environ 28:233–249

Cabrera MR (1998) Las Tortugas Continentales de Sudamérica Austral. Talleres gráficos BR Copias, Córdoba

Caccone A, Amato G, Gratry OC, Behler J, Powell JR (1999a) A molecular phylogeny of four endangered Madagascar tortoises based on mtDNA sequences. Mol Phylogenet Evol 12:1–9

Caccone A, Gibbs JP, Ketmaier V, Suatoni E, Powell JR (1999b) Origin and evolutionary relationships of giant Galápagos tortoises. Proc Natl Acad Sci USA 96:13223–13228

Caccone A, Gentile G, Gibbs JP, Fritts TH, Snell HL, Betts J, Powell JR (2002) Phylogeography and history of giant Galápagos tortoises. Evolution 56:2052–2066

Cadena E (2009) New Cretaceous and Cenozoic Fossil turtles from Colombia and Panama; systematic paleontology, phylogenetical and paleobiogeographical implications. Dissertation MSc, University of Florida, Gainesville

Cadena EA, Bloch JI, Jaramillo CA (2010) New podocnemidid turtle (Testudines: Pleurodira) from the Middle-Upper Paleocene of South America. J Vert Paleontol 30:367–382

Cadena E, Bourque JR, Rincon A, Bloch JI, Jaramillo CA, MacFadden BI (2012) New turtles (Chelonia) from the Late Eocene through Late Miocene of the Canal Basin. J Paleontol 86:539–557

Castaño Mora OV, Lugo M (1981) Estudio comparativo del comportamiento de dos especies de morrocoy: *Geoechelone carbonaria* y *Geoehelone denticulata* y aspectos com parables de su morfología externa. Cespedesia 10:55–122

Cei JM (1986) Reptiles del centro, centro-oeste y sur de laArgentina. Herpetofauna de las zonas áridas y semiáridas. In: Museo Regionale di Scienze naturali, Monogr IV, Torino

Chkhikvadze VM (1972) Sur la position systemátique des Tortue terrestres géantes du Tertiaire du Paléoarctique. Soovshch Akad Nauk Guznuskoj 65:745–748

Claude J, Tong H (2004) Early Eocene testudinoid turtles from Saint Papoul, France, with comments on the early evolution of modern Testudinoidea. Oryctos 5:3–45

Cope ED (1872) Description of some new Vertebrata from the Bridger group of the Eocene. Paleontol Bull 1:1–60

Crumly CR (1982) A cladistic analysis of *Geochelone* using cranial osteology. J Herpetol 16:215–234

Crumly CR (1984a) The evolution of land tortoises (family Testudinidae). PhD dissertation, Rutgers University, Newark

Crumly CR (1984b) The cranial morphometry of Galapagos tortoises. Proc Calif Acad Sci 43:111–121
Crumly CR (1984c) A hypothesis for the relationships of land tortoises genera (Family Testudinidae). Stud Geol Salmant Vol Espec 1:115–124
de Broin F (1991) Fossil turtles form Bolivia. In: Suarez-Soruco R (ed) Fósiles y Facies de Bolivia, vol 1 Vertebrados, vol 12, Rev Téc Yac Petrolíf Fis Bol. YPFB, Santa Cruz, pp 509–527
de la Fuente MS (1988) Las tortugas Chelidae (Pleurodira) y Testudinidae (Cryptodira) del Cenozoico argentino. Doctoral dissertation, Universidad Nacional de La Plata
de la Fuente MS (1994) Descripción de nuevos especímenes y relaciones filogenéticas de *Chelonoidis gringorum* (Chelonii: Testudinidae) del Mioceno temprano de Patagonia central. Stu Geol Salmant 30:107–131
de la Fuente MS (1997a) Las tortugas terrestres gigantes del Mioceno tardío-Plioceno del territorio argentino. Stud Geol Salmant 33:91–120
de la Fuente MS (1997b) Las tortugas pleistocenas del extremo meridional de la provinciade Santa Fé, Argentina. Stu Geol Salmant 33:67–90
de la Fuente MS (1999) A review of the Pleistocene Reptiles of Argentina: taxonomic and palaeoenvironmental considerations. Quat South Am Antar Penin 12:111–138
de la Fuente MS, Cabrera M (1988) Comentarios sobre las tortugas fósiles de Sierra Chica (provincia de Córdoba, Argentina). Bol Asoc Herpetol Arg 4:65–74
Duméril AMC, Bibron G (1834) Erpétologie générale au histoire naturelle complète des reptiles, vol I. Librairie Encyclopédique de Roret, Paris
Ernst CH, Barbour RW (1989) Turtles of the world. Smithsonian Institution, Washington, DC
Ernst CH, Altenburg RGM, Barbour RW (2000) Turtles of the World. World Biodiversity Database, CD-ROMSeries, Windows, Version 1.2. Biodiversity Center of ETI, Amsterdam
Fernández MS (1988) Las Testudinidae (Reptilia: Chelonii) argentinas: Osteología, sistemática y distribución geográfica. Doctoral dissertation, Universidad Nacional de La Plata, La Plata
Fitzinger L (1835) Entwurf einer systematischen Anordnung der Schildkröten nach den Gründsatzen der natürlichen Methode. Ann Wien Mus Natu 1:105–128
Freiberg MA (1973) Dos nuevas tortugas terrestres de Argentina. Bol Soc Biol Concepción 46:81–93
Frits U, Alcalde L, Vargas Ramírez M, Goode EV, Fabious-Turoblin DU, Praschag P (2012) Northern genetic richness and southern purity, but just one species in *Chelonoidis* complex. Zool Scripta 41:220–232
Fritts T (1983) Morphometrics of Galapagos tortoises: evolutionary implications. In: Bowman R, Berson M, Leviton AE (eds) Patterns of evolution in Galapagos organisms, Am Assoc Advance Sci Pac Div. Pacific Division, AAAS, San Francisco
Fritz U, Bininda-Emonds ORP (2007) When genes meet nomenclature: tortoise phylogeny and the shifting generic concepts of *Testudo* and *Geochelone*. Zool 110:298–307
Fritz U, Siroky P, Kami H, Wink M (2005) Environmentally caused dwarfism or a valid species – is Testudo weissingeri Bour 1996 a distinct evolutionary lineage? New evidence from mitochondrial and nuclear genomic markers. Mol Phylogenet Evol 37:389–401
Gaffney ES, Meylan PA (1988) A phylogeny of turtles. In: Benton MJ (ed) The phylogeny and classification of the tetrapods. 1: Amphibians, reptiles, birds, vol 35A, Systematics association, special volume. Clarendon, Oxford
Garman S (1917) The Galapagos tortoises. Mem Mus Comp Zool 30:261–296
Gerlach J (2001) Tortoise phylogeny and the '*Geochelone*' problem. Phelsuma 9:1–24
Gray JE (1854) Description of a new genus and some new species of tortoises. Proc Zool Soc Lond 1854:133–135
Gray JE (1870) Notice of a new Chilean tortoise (*Testudo chilensis*). Ann Mag Nat Hist Lond 6:190–191
Gray JE (1873) Hand-list of the specimens of shield reptiles in the British Museum. British Museum, London

Günther ACLG (1875) Description of the living and extinct races o gigantic land tortoises. Part 1–2. Introduction and the tortoises of the Galapagos Island. Philos Trans R Soc Lond Biol Sci 165:251–284

Günther ACLG (1877) The gigantic land tortoises (living and extinct). In: The collection of the British Museum. Taylor & Francis, London

Harlam R (1827) Description of a land tortoise, from the Galapagos Islands, commonly know as the "Elephant Tortoise". J Acad Nat Sci Phila 5:284–292

Head JJ, Aguilera OA, Sánchez Villagra M (2006) Past colonization of South America by trionychid turtles: fossil evidence from the Neogene of Margarita Island, Venezuela. J Herpetol 40:378–381

Hoffstetter R (1972) Relationships, origins and phylogeny of the ceboid monkeys and caviomorph rodents: a modern interpretation. Evol Biol 6:323–347

Hoffstetter R (1974) Phylogeny and geographic development of Primates. J Human Evol 3:323–350

Hoffstetter R (1975) El origen de los caviomotpha y el problema de los hystricognathi (Rodentia). Actas I Congr Argent Paleont Bioestr 2:505–528

Hoffstetter R (1980) Los monos platirrinos (Primates): Origen, extensión, filogenia, taxonomía. Actas II Congr Argent Paleont Bioestr I Congr Latinoamer Paleont 2:291–303

Hoffstetter R, Lavocat R (1970) Découverte dans le Déséadien de Bolivie de genres pentaloph-odontes appuyant les affinités africaines dês rongeurs caviomorphes. C R Acad Sci Paris 271:172–175

Holroyd PA, Parham JF (2003) The antiquity of African tortoises. J Vertebr Paleontol 23:688–690

Hutchison JH (1996) Testudines. In: Prothero DR, Emry RJ (eds) The terrestrial Eocene-Oligocene transition in North America. Cambridge University Press, New York

Hutchison JH (1998) Turtles across the Paleocene/Eocene epoch boundary in West-Central North America. In: Aubry MP, Lucas S, Berggren WA (eds) Late Paleocene-early Eocene climate and biotic events in the marine and terrestrial records. Columbia University Press, New York

Iriondo M (1996) Estratigrafía del Cuaternario de la cuenca del río Uruguay. 13 Congr Geol Arg 3 Congr Expl Hidrocar Actas 4:15–25

Iriondo M, Cerutti C, Tardivo R (1985) Geomorfología y Cuaternario del tramo inferior del Arroyo Feliciano. Rev Asoc Cien Nat Litoral 16:149–156

Iverson JB (1992) A revised checklist with distribution maps of the turtles of the world. Privately Printed, Richmond

Iverson JB, Spinks PQ, Shffer HB, McCord WP, Das I (2001) Phylogenetic relationships among the Asian tortoises of the genus Indotestudo (Reptilia: Testudines: Testudinidae). Hamadryad 26:272–275

Joyce WG, Parham JF, Gauthier JA (2004) Developing a protocol for the conversion of rank-based taxon names to phylogenetically defined clade names, as exemplified by turtles. J Paleontol 78:989–1013

Khosatzky LI, Mlynarsky M (1966) *Agrionemys* nouveau genre de tortues terrestres (Testudinidae). Bull Acad Po Sci 14:123–125

King FW, Burke RL (1989) Crocodilian, tuatara and turtle species on the world. Association of Systematics Collections, Washington, DC

Krenz JG, Naylor GJP, Shaffer HB, Janzen FJ (2005) Molecular phylogenetics and evolution of turtles. Mol Phylogenet Evol 37:178–191

Lamb T, Lydeard C (1994) A molecular phylogeny of the Gopher tortoises, with comments on familial relationships within the Testudinoidea, with comments on familial relationships within the Testudinoidea. Mol Phylogenet Evol 3:283–291

Lapparent de Broin F (2000) Les Chéloniens de Sansan. In: Ginsburg L (ed) La fauna miocène de Sansan et son environnement, Mém Mus Nt Hist Nat Paris, vol 183, pp 219–261

Laurent R (1979) Herpetofaunal relationships between Africa and South America. In: Duellman W (ed) The South American Herpetofauna. Its origins, evolution and dispersal, Mus Nat Hist Univ Kansas Monogr, vol 7, pp 55–72

Lavocat R (1969) La systématique des Rongers histricomorphes et la dérive des continents. C R Acad Sci Paris 269:1496–1497
Lavocat R (1977) Sur l'origine des faunas sud-américaines de mammifères du Mésozoique terminal et du Cénozoic ancient. C R Acad Sci Paris 285:1423–1426
Le M, McCord W (2008) Phylogenetic relationships and biogeographical history of the genus *Rhinoclemmys* Fitzinger, 1835 and the monophyly of the turtle family Geoemydidae (Testudines: Testudinoidea). Zool J Linn Soc 153:751–767
Le M, Raxworthy CJ, McCord WP, Mertz L (2006) A molecular phylogeny of tortoises (Testudines: Testudinidae) based on mitochondrial and nuclear genes. Mol Phylogenet Evol 40:517–531
Leidy J (1851) On a species of fossil tortoise. Proc Acad Nat Sci Phila 5:172–173
Lindholm WA (1929) Revidierttes Verzeichnis der Gattungen der rezenten Schildkröten nebst Notizen zur Norre1klatur einiger Arten. Zool Anz 81:275–295
Linnaeus C (1758) Systema Naturae per regna tria naturae, secundum classes, ordines, genera, species, cum characteribus, differentiis, synonymis, locis. Editio decima, reformata. Tomus I. Laurentii Salvii, Holmiae
Linnaeus C (1766) Systema Naturae. Editio duodecima. Halae Magdeboricae 1:1–532
Loveridge A, Williams E (1957) Revision of the African tortoises and turtles of the suborder Cryptodira. Bull Mus Comp Zool 115:163–557
Manzano A, Noriega JI, Joyce W (2009) The tropical tortoise *Chelonoidis denticulata* (Testudines: Testudinidae) from the late Pleistocene of Argentina and its paleoclimatological implications. J Paleontol 83:975–980
Marlow RW, Patton J (1981) Biochemical relationships of the Galápagos Giant tortoise (*Geochelone elephantopus*). J Zool Soc Lond 195:413–422
Marshall LG, Patterson B (1981) Geology and Geochronology of the mammal bearing Tertiary of the Valley of Santa María and Río Corral Quemado, Catamarca Province, Argentina. Field Geol 9:1–80
Mendía J, Bayarsky A (1981) Estratigrafía del Terciario en el Valle inferior del río Chubut. Actas 7 Congr Geol Arg 3:593–606
Meylan PA, Auffenberg W (1986) New land tortoises (Testudines: Testudinidae) from the Miocene of Africa. Zool J Linn Soc 86:279–307
Meylan PA, Sterrer W (2000) *Herperotestudo* (Testudines: Testudinidae) from the Pleistocene of Bermuda, with comments on the phylogenetic position of the genus. Zool J Linn Soc 128:51–76
Moreno FP (1889) Breve reseña de los progresos del Museo de la Plata, durante el segundo semestre de 1888. Bol Mus La Plata 2:1–44
Near TJ, Meylan PA, Shaffer HB (2005) Assessing concordance of fossil calibration points in molecular clock studies: an example using turtles. Am Nat 165:137–146
Palkovacs EP, Gerlach J, Caccone A (2002) The evolutionary origin of Indian Ocean tortoises (*Dipsochelys*). Mol Phylogenet Evol 24:216–227
Parham JF, Macey JR, Papenfuss TJ, Feldman CR, Turkozan O, Polymeni R, Boore J (2006) The phylogeny of Mediterranean tortoises and their close relatives based on complete mitochondrial genome sequences from museum specimens. Mol Phylogenet Evol 38:50–64
Pascual R (1984a) La sucesión de las edades-mamíferos, de los climas y del diastrofismo sudamericano durante el Cenozoico: fenómenos concurrentes. An Acad Nac Cs Ex Fis Nat Buenos Aires 36:15–37
Pascual R (1984b) Late Tertiary mammals of southern South America as indicators of climatic deterioration. Quat South Am Antarc Pen 1:1–30
Patterson B, Pascual R (1972) The fossil mammals fauna of South America. In: Keast A, Erk FC, Glass B (eds) Evolution of mammals and southern continents. State University New York Press, Albany
Pictet FJ, Humbert A (1855–1857) Ordre des Chéloniens. In: Pictet FJ, Gaudin CT, De La Harpe (eds) Mémoire sur le animaux vertébrés trouvés dans le terrain sidérolithique du canton de Vaud et appartenant à la faune Eocène, Genève

Pritchard PCH (1984) Evolution and zoogeography of south American turtles. Stud Geol Salmant Vol Espec 1:225–233

Pritchard PCH (1996) The Galápagos Tortoises. Nomenclatural and survival status. Chel Res Monogr 1:7–85

Pritchard PCH, Trebbau P (1984) The turtles of Venezuela. Contrib Herpetol 2:1–403

Quay JR, Gaimard JP (1824) Sous-gene Tortue de Terre *Testudo nigra*. In: Freycinet ML (ed) Voyage autor du monde execute sur L'Ucranie et la Physicienne pendant les années 1817–1820, Paris

Rafinesque CS (1832) Description of two new genera of turtles of North America. Atl J Friend Knowl Phila 1:64–65

Reig O (1962) Las integraciones cenogenéticas en el desarrollo de la fauna de vertebrados tetrápodos de América del Sur. Ameghiniana 1:131–140

Reig O (1968) Peuplement en vértébres tétrapodes de l'Amerique do Sud. In: Biologie de la Amérique Austral Delamare, Debouterville CI, Rapopport E, CNRS, Paris

Reig O (1981) Teoría del origen y desarrollo de la fauna de mamíferos de América del Sur. Monogr Nat 1:1–162

Rhodin AGJ, Van Dijk PP, Iverson JB, Broadley Shaffer J (2010) Turtles of the world, 2010 update: annoted checklist of taxonomy, synonymy, distribution, and conservation status. In: Rhodin AGJ, Pritchard PCH, van Djik PP, Saumure RA, Buhlman KA, Iverson JB, Mittermeier RA (eds) Conservation biology of freshwater turtles and tortoises: a compilation project of the IUCN/SSC tortoises and freshwater turtle specialist group, vol 5, Chelonian research monographs., pp 000.85–000.164

Richard E (1999) Tortugas de las regiones áridas de Argentina. Contribución al conocimiento de las tortugas áridas de Argentina (Chelidae y Testudinidae) con especial referencia a los aspectos ecoetológicos, comerciales y antropológicos de las especies de complejo *chilensis* (*Chelonoidis chilensis* y *C. donosobarrosi*) enla provincia de Mendoza. LO L A (Literature of Latin America), Buenos Aires

Riggs E, Patterson B (1939) Stratigraphy of the Late Miocene deposits of the Late Miocene deposits of the province of Catamarca (Argentina) with notes on the faunae. Physis 14:143–162

Rothschild W (1901) On a new land tortoise from the Galapagos Island. Nov Zool 8:372

Rothschild W (1903) Description of a new species of gigantic land tortoise from Indefatigable Island. Nov Zool 10:119

Rovereto C (1914) Los Estratos Araucanos y sus fósiles. An Mus Hist Nat Buenos Aires 25:1–24

Russello MA, Glaberman S, Gibbs JP, Marquez C, Powell JR, Caccone A (2005) A cryptic taxon of Galápagos tortoise in conservation peril. Biol Lett 1:287–290

Sánchez Villagra MR, Asher RJ, Rincon AD, Carlini AA, Meylan PA, Purdy RA (2004) New faunal reports for the Cerro La Cruz locality (Lower Miocene), north western Venezuela. In: Sánchez Villagra MR, Clark JA (eds) Fossil of the Miocene Castillo formation, Venezuela: contributions on neotropical palaeontology, Spec Pap Palaeontol, vol 71, pp 105–112

Simons E (1976) The fossil record of primate phylogeny. In: Tashian R (ed) Molecular anthropology Goodman M. Academic, New York

Simpson GG (1942) A Miocene tortoise from Patagonia. Am Mus Novit 1209:1–6

Simpson GG (1943) Turtles and the origin of the fauna of Latin America. Am J Sci 241: 413–429

Simpson GG (1980) Splendid isolation: the curious history of South American mammals. Yale University Press, New Haven

Spinks PQ, Shaffer HB, Iverson JB, McCord WP (2004) Phylogenetic hypotheses for the turtle family Geoemydidae. Mol Phylogenet Evol 32:164–182

Spix J B von (1824) Animalia nova. Species novae Testudinum et Ranarum quas in itinere per Brasiliam ann is 1817–1820 collegit et descripsit. Hubschmann, München

Sukhanov VB (2000) Mesozoic turtles of Middle and Central Asia. In: Benton MJ, Shishkin MA, Unwin DM, Kurochkin EN (eds) The age of dinosaurs in Russia and Mongolia. Cambridge University Press, Cambridge

Tarling D (1980) The geological evolution of South America with special reference to the last 200 million years. In: Ciochon RL, Chiarelli AB (eds) Evolutionary biology of the New Worlds monkeys and continental drift. Plenum Press, New York

Van Denburgh J (1907) Preliminary description of four races of land tortoises from Galapagos Island. Proc Calif Acad Sci 4:1–6

van der Kuyl AC, Ballasina DLP, Dekker JT, Maas J, Willemsen RE, Goudsmit J (2002) Phylogenetic relationships among the species of the genus Testudo (Testudines: Testudinidae) inferred from mitochondrial 12S rRNA gene sequences. Mol Phylogenet Evol 22:174–183

Williams EE (1950) *Testudo cubensis* and the evolution of western hemisphere tortoises. Bull Am Mus Nat Hist 95:1–36

Williams EE (1952) A new fossil tortoise from Mona Island, West Indies and a tentative arrangement of the tortoises of the world. Bull Am Mus Nat Hist 9:541–560

Williams EE (1960) Two species of tortoises in northern South America. Breviora 120:1–13

Zacarias G, de la Fuente MS, Fernández M, Zurita A (in press) Nueva especie de tortuga terrestre gigante del género *Chelonoidis* Fitzinger 1835 (Cryptodira: Testudinidae), de la Provincia de Corrientes, Argentina. Ameghiniana, Buenos Aires.

Chapter 4
South Gondwana Pleurodiran Turtles

Abstract Extant and extinct species of chelid turtles are limited in distribution to South America and Australasia with an extensive fossil record in South America (Early Cretaceous–Recent) and more scarce in Australasia (Early Cretaceous, Eocene, Oligocene, and Miocene to Recent). Taking into account their specific richness, chelids are one of the most diverse South American turtle clades. This fact and the above-mentioned disjointed distribution, the early fossil record, and the phylogenetic data suggest an early differentiation from Pelomedusoides at least in the Early Cretaceous in South Gondwanan landmasses.

Keywords Chelidae • Gondwana • Cretaceous • Cenozoic • South America • Australasia

4.1 Chelidae Lindholm 1929

Chelidae is a group of freshwater aquatic to semiaquatic eupleurodiran turtles with extant species distributed in South America and Australasia (Pritchard and Trebbau 1984; Iverson 1992) (Fig. 4.1a), the fossil record of which is known as early as in the Aptian–Albian of Patagonia (Lapparent de Broin and de la Fuente 2001; de la Fuente et al. 2011) and the Albian of Australia (Smith 2009, 2010). This disjointed geographic range of extant members and the early fossil record of this clade might be the result of a biogeographic and phylogenetic history developed on Gondwanan landmasses. These turtles are characterized by two main autapomorphic conditions: (1) the formula of articulation of the cervical vertebrae includes, adjacent to the biconvex atlas, three opisthocoelous, two biconvex, one procoelous, one biconcave vertebrae, and one biconvex [=)1(, (2(, (3(, (4(, (5),)6),)7(, (8); after Williams (1950) notation] and (2) the development of a lateral cheek emargination with the loss of the quadratojugal (Gaffney 1977; Broin and de la Fuente 1993).

M.S. de la Fuente et al., *Origin, Evolution and Biogeographic History of South American Turtles*, Springer Earth System Sciences, DOI 10.1007/978-3-319-00518-8_4,

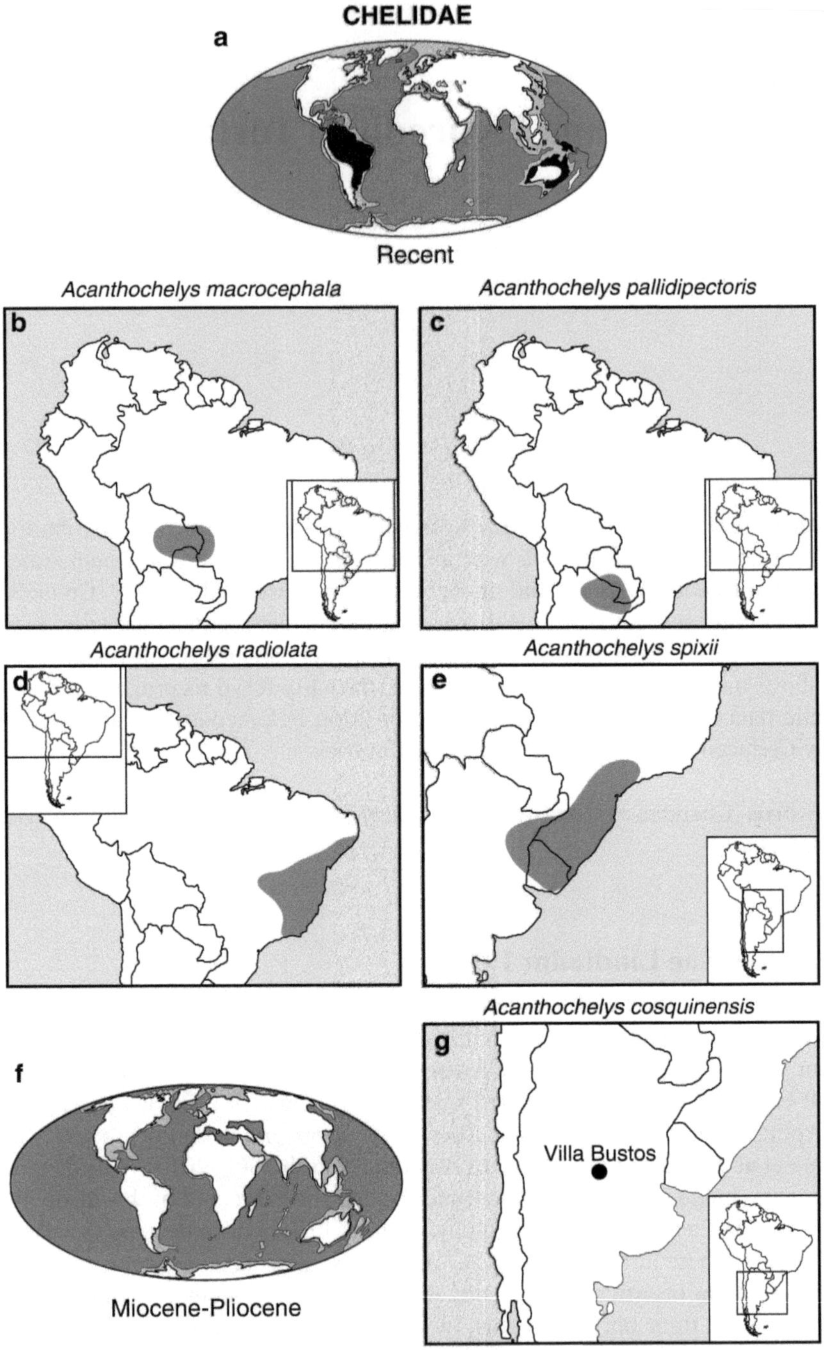

Fig. 4.1 Geographic distribution of Chelidae in recent times. (**a**) Global distribution of Chelidae. (**b**) Geographic distribution of *Acanthochelys macrocephala* in South America. (**c**) Geographic distribution of *Acanthochelys pallidipectoris* in South America. (**d**) Geographic distribution of *Acanthochelys radiolata* in South America. (**e**) Geographic distribution of *Acanthochelys spixii* in South America. (**f**) Paleogeographic reconstruction of continents during the Miocene–Pliocene. (**g**) Geographic and stratigraphic location of *Acanthochelys cosquinensis* in South America

Genera included: *Acanthochelys* Gray 1873, *Bonapartemys* Lapparent de Broin and de la Fuente 2001, *Chelodina* Fitzinger 1826, *Chelus* Duméril 1805, *Elseya* Gray 1867, *Elusor* Cann and Legler 1994, *Emydura* Bonaparte 1836, *Hydromedusa* Wagler 1830, *Linderochelys* de la Fuente, Calvo and González Riga 2007, *Lomalatachelys* Lapparent de Broin and de la Fuente 2001, *Mesoclemmys* Gray 1873, *Myuchelys* Thomson and Georges 2009, *Palaeophrynops* Lapparent de Broin and de la Fuente 2001, *Phrynops* Wagler 1830, *Platemys* Wagler 1830, *Prochelidella* Lapparent de Broin and de la Fuente 2001, *Pseudemydura* Siebenrock 1901, *Rheodytes* Legler and Cann 1980, *Rhinemys* Wagler 1830, *Salamanchelys* Bona 2006, and *Yaminuechelys* de la Fuente, Lapparent de Broin and Manera de Bianco 2001.

South American Taxa

4.1.1 *Acanthochelys* Gray 1873

Type species: *Emys radiolata* Mikan 1820.
Species included: *Acanthochelys cosquinensis* de la Fuente 1992, *A. macrocephala* (Rhodin, Mittermeier and McMorris 1984), *A. pallidipectoris* (Freiberg 1945), *A. radiolata* (Mikan 1820), and *A. spixii* (Duméril and Bibron 1835).

South American Species

Acanthochelys macrocephala (Rhodin, Mittermeier and McMorris 1984)

Holotype: Naturhistorisches Museum Wien, Austria. NMW 1293.
Diagnosis: See Rhodin et al. (1984).
Distribution: This species is known from the upper Río Mamore drainage of central Bolivia and the Pantanal region and other swamplands of the upper Río Paraguay drainage in southwestern Mato Grosso, Brazil (Rhodin et al. 1984, 2009) (Fig. 4.1b).
Comments: This species was first named and assigned by Rhodin et al. (1984) to the genus *Platemys* Wagler 1830 as *Pl. macrocephala*. However, as noted by Derr et al. (1987) biochemical, karyological, and morphological differences exist between *Platemys platycephala* and the other species at that time referred to *Platemys*. These differences prompted Derr et al. (1987) to suggest the recognition of two different generic entities. Consequently, with this taxonomic conclusion, Iverson (1986) resurrected the genus *Acanthochelys* Gray 1873 and referred the four other extant species currently assigned to *Platemys* (*macrocephala*, *pallidipectoris*, *radiolata*, and *spixii*) to the genus *Acanthochelys*. A detailed external morphological and osteological description of *Acanthochelys macrocephala* is presented by Rhodin et al. (1984).

Although Rhodin et al. (1984) have suggested that *A. macrocephala* is more closely related to *A. radiolata* than the remaining extant *Acanthochelys* species, the cytogenetic and morphological data used by these authors to differentiate these species might also allow the differentiation of *A. macrocephala* from the remaining extant *Acanthochelys* species (e.g., *A. pallidipectoris, A. radiolata, A. spixii*). Among these defining characters, Rhodin et al. (1984) included head width, skull and carapace height, postorbital bar width, interorbital width, supraoccipital width, degree of divergence of pterygoid processes, and maxillar and mandibular surface width. The authors also provided brief notes on the ecology, reproduction, and diet of this species, showing the major differences in morphology, osteology, and karyotypes and the allopatric geographic range between *A. macrocephala* and *A. radiolata*. More recently, updated information concerning taxonomy, distribution, ecology, and conservation of the big-head Pantanal swamp turtle is summarized by Rhodin et al. (2009).

Acanthochelys pallidipectoris (Freiberg 1945)

Holotype: Museo Argentino de Ciencias Naturales "Bernardino Rivadavia" Buenos Aires, Argentina. MACN 1731.

Definition: See Ernst (1983a).

Distribution: Chaco region of northern Argentina, southeastern Bolivia, and western Paraguay (Vinke et al. 2011, and references therein) (Fig. 4.1c).

Comments: Freiberg (1945) named this species *Platemys pallidipectoris* based on a specimen from Presidencia Roque Saenz Peña, Chaco Province of northern Argentina. Based on McBee et al. (1985) karyological study and Rhodin (1985) taxonomic suggestion, *pallidipectoris* is assigned to the genus *Acanthochelys* Gray 1873 by Iverson (1986). This species was reviewed by Ernst (1983a) and Cabrera (1998), and the latter provided a detailed description and figures of the skull and shell. A detailed summary of this species and a revised assessment of its habitat, ecology, population, and conservation are presented by Vinke et al. (2011).

Acanthochelys radiolata (Mikan 1820)

Holotype: Naturhistorisches Museum Wien, Austria. NMW 23390.

Definition: See Ernst (1983c).

Distribution: Southeastern Brazil, Atlantic coastal drainages from the San Francisco river to Sao Paulo State, and possibly the upper Xingu River basin and Mato Grosso (King and Burke 1989; Iverson 1992) (Fig. 4.1d).

Comments: The Brazilian radiolated swamp turtle is endemic to regions of the Atlantic Brazilian Forest. It was first described by Mikan (1820) as *Emys radiolata* and later reviewed by Ernst (1983c) as *Platemys radiolata*. Ernst (1983c) and Ernst and Barbour (1989) provided a brief description of this species. The reproductive biology and natural history in captivity are updated by Mocelin et al. (2008).

Acanthochelys spixii (Duméril and Bibron 1835)

Holotype: Muséum National d'Histoire Naturelle de Paris, France. MNHN 8751.
Definition: See Ernst (1983b).
Distribution: This species occurs in southeastern Brazil from the San Francisco River southward to the coastal drainages of Uruguay and Corrientes Province, Argentina (Iverson 1992) (Fig. 4.1e).
Comments: This species was described as *Emys depressa* by Spix (1824). Unfortunately, this name was not available because it had been used previously by Merrem (1820) to describe other chelid specimens. More recently, when Duméril and Bibron (1835) described MNHN 8751, they referred to it as *Platemys spixii,* and it has since become the holotype of this taxon (Rhodin et al. 1984). Likewise, one of Spix's syntypes of *Emys depressa* ZSM was posteriorly designed by Hoogmoed and Grüber (1983) as the lectotype of this species. Spix's side-necked turtle was reviewed by Groombridge (1982), Ernst (1983b), and Cabrera (1998). The latter author provided a detailed description of this species and figures of the skull and shell.

Acanthochelys cosquinensis de la Fuente 1992

Holotype: Museo de La Plata, La Plata, Argentina. MLP 77-V-12-1. Carapace, plastron, two cervical vertebrae (fourth, sixth or seventh), first dorsal vertebra, and left humerus.
Diagnosis: See de la Fuente (1992).
Locality, horizon, and age: Reddish brown clay sediment of an unnamed formation assigned to the Montehermosan–Chapadmalalan SALMAs (Pliocene) (see de la Fuente 1992; Cione and Tonni 1995) outcropping at the Cosquín River between the villages of Cosquín and Santa María, near Villa Bustos, Córdoba Province, Argentina (de la Fuente 1986, 1992) (Fig. 4.1f, g).
Comments: The first reference to chelids in the Punilla Valley was presented by de la Fuente and Ledesma (1985) who, in a brief note, assigned the specimen MLP 77-V-12-1 to the genus *Platemys* Wagler 1830. More recently, this specimen was described by de la Fuente (1986). Later, de la Fuente (1988) and de la Fuente and Cabrera (1988) assigned this turtle to the genus *Platemys* (*sensu lato*) in accordance with McBee et al. (1985). These authors pointed out that *Platemys*, as presently defined, shows a greater range of variation than occurs in any other member of this clade of crown Pleurodira. The karyotypic pattern reported by McBee et al. (1985) indicates removal of all species except *Pl. platycephala* from the genus *Platemys*. This distinction is also supported by biochemical data presented by Derr et al. (1987). The remaining four extant species are assigned to the genus *Acanthochelys* Gray 1873 (see details above). Finally, de la Fuente (1992) assigned the Punilla Valley chelid turtle to a new species of *Acanthochelys* (*A. cosquinensis*). According to de la Fuente (1992), the first extinct *Acanthochelys* species is more closely related to the extant *A. radiolata* than to any other *Acanthochelys* species.

4.1.2 *Prochelidella* Lapparent de Broin and de la Fuente 2001

Type species: *Prochelidella argentinae* Lapparent de Broin and de la Fuente 2001 (Lapparent de Broin and de la Fuente 2001, Fig. 3).
Species included: *Prochelidella argentinae* Lapparent de Broin and de la Fuente 2001, *P. cerrobarcinae* de la Fuente, Umazano, Sterli and Carballido 2011, and *P. portezuelae* de la Fuente 2003.

Prochelidella argentinae Lapparent de Broin and de la Fuente 2001

Holotype: Museo Argentino de Ciencias Naturales "Bernardino Rivadavia," Buenos Aires, Argentina. MACN-CH 1680, anterior part of the carapace.
Diagnosis: See Lapparent de Broin and de la Fuente (2001) and de la Fuente (2007).
Locality and age: Located 6 km north of "Estancia Ocho Hermanos," Sierra de San Bernardo, Sarmiento Department, Chubut Province, Argentina. Upper part of the Lower Member of the Bajo Barreal Formation, Upper Cretaceous, Turonian–Campanian (Martínez et al. 1986; Barcat et al. 1989, Leanza 1999), or Cenomanian–Turonian (Bridge et al. 2000) (Fig. 4.2a, b).
Comments: This species[1] was originally defined by Lapparent de Broin and de la Fuente (2001: 466–467, Fig. 3) on the basis of a specimen consisting of the anterior margin of a low, wide carapace. These authors also suggested that the specimen might be related to extant species of the genus *Acanthochelys* (Gray 1873) on account of its small size and the decoration characterized by a dense microvermiculation with rounded ridges. However, this species retains plesiomorphic traits such as a wide and short nuchal bone and cervical scute, the presence of neurals, and more advanced axillar process, which, as suggested by Lapparent de Broin and de la Fuente (2001), distinguished it from the extant species of the genus. Lapparent de Broin and de la Fuente (2001) reported additional remains of small forms similar or very close to *Prochelidella argentinae* from Lower and Upper Cretaceous sites of Neuquén and Río Negro Provinces (Patagonia, Argentina). These specimens (most of them represented by isolated shell fragments) indicate the presence of several forms belonging to *Prochelidella*. Some carapace and plastral traits of these specimens (Broin and de la Fuente 1993) including short pygals (rectangular or trapezoid), pygal bone well overlapped by vertebral 5, the pygal posterior border transverse or slightly notched or rounded, and the plastral scute pattern support this assignment.

Prochelidella portezuelae de la Fuente 2003

Holotype: Museo Carmen Funes Plaza Huincul, Neuquén Province, Argentina. MCP-PVPH-161, anterior margin of the carapace and nearly complete plastron, left atlantal arch and five other cervical vertebrae, both girdles, and left and right centrodistal extremity of the right femur and right tibia.

[1]Courtesy of Indiana University Press.

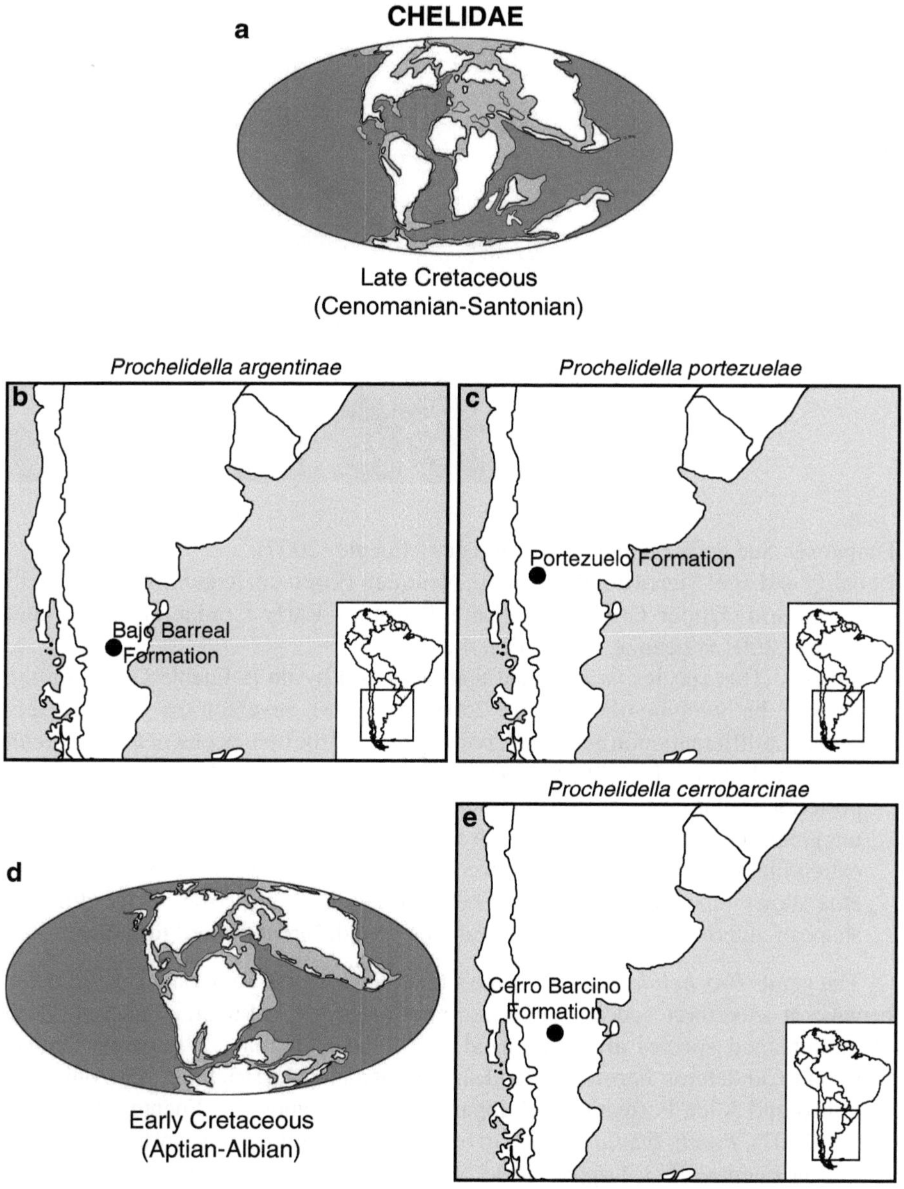

Fig. 4.2 South American record of *Prochelidella* in the Cretaceous. (**a**) Paleogeographic reconstruction of continents during the Late Cretaceous. (**b**) Geographic and stratigraphic location of *Prochelidella argentinae*. (**c**) Geographic and stratigraphic location of *Prochelidella portezuelae*. (**d**) Paleogeographic reconstruction of continents during the Early Cretaceous. (**e**) Geographic and stratigraphic location of *Prochelidella cerrobarcinae*

Fig. 4.3 Life reconstruction of *Prochelidella cerrobarcinae* from the Early Cretaceous of the Cerro Barcino Formation, Argentina

Diagnosis: See de la Fuente (2003) and de la Fuente (2007).

Locality and age: Sierra del Portezuelo, Neuquén Province, Argentina. Portezuelo Formation, Upper Cretaceous (Late Turonian–Early Coniacian; Hugo and Leanza 2001a; Leanza 1999) (Fig. 4.2a, c).

Comments: This species[2] was named and described by de la Fuente (2003) on the basis of the holotype (de la Fuente 2003, Figs. 1–8). Several traits suggest a species-level differentiation between *P. portezuelae* and the two species of *Prochelidella* mentioned above. In contrast to *P. argentinae*, *Pr. portezuelae* includes a more posterior axillar process, a distinct outline and proportions of the first and second marginals, absence of marked growth annuli, the moderate size, and week free ribs extremities of pleural 1. Likewise, in comparison with *P. cerrobarcinae*, *P. portezuelae* shows moderate size, contact between neural 2 and costal 1, the different shape of anterior and posterior plastral lobe, and different plastral formulae.

The genus *Prochelidella* has become a relatively diverse Patagonian chelid taxon because at least three species have been nominated (de la Fuente et al. 2011) and other unnamed species are recognized in Albian–Cenomanian horizons (Lohan Cura and Candeleros Formation) and in Campanian–Maastrichtian horizons (Los Alamitos and Allen Formations) (Lapparent de Broin and de la Fuente 2001; de la Fuente 2007). *Prochelidella* appears to be the oldest chelid genus worldwide (Broin and de la Fuente 1993; Lapparent de Broin and de la Fuente 2001; de la Fuente 2007; de la Fuente et al. 2011).

Prochelidella cerrobarcinae de la Fuente, Umazano, Sterli, Carballido 2011 (Fig. 4.3)

Holotype: Museo Paleontológico Egidio Feruglio, Trelew, Argentina. MPEF-PV 3287, the cast and bones of the anterior margin of the carapace.

[2]Courtesy of Indiana University Press.

Diagnosis: See de la Fuente et al. (2011).

Locality and age: Los Chivos Hill. Puesto La Paloma Member of the Cerro Barcino Formation, Chubut Group. Aptian–Albian? in age (Volkheimer et al. 2009; de la Fuente et al. 2011) (Fig. 4.2d, e).

Comments: *Prochelidella cerrobarcinae* was named by de la Fuente et al. (2011) on the basis of the holotype (MPEF-PV-3287) and 10 referred specimens (MPEF-PV 3288–3297), mostly known through shells and, in smaller proportion, by remains of appendicular skeletons. The morphology of the anterior part of the carapace of *Prochelidella cerrobarcinae* (MPEF-PV-3287) is consistent with the diagnostic characteristics of the genus *Prochelidella* (Lapparent de Broin and de la Fuente 2001): a low, wide carapace with a slight nuchal notch, moderate elongation of the anterior border of the carapace, a nuchal bone anteriorly and posteriorly widened, and a rectangular neural 1 is recognized in the holotype of *P. cerrobarcinae* (Lapparent de Broin and de la Fuente 2001; de la Fuente 2003; de la Fuente et al. 2011). These authors have suggested that the three named species of *Prochelidella* (*P. argentinae*, *P. portezuelae*, and *P. cerrobarcinae*) might be related to extant species of the genus *Acanthochelys* (Gray 1873). However, this hypothesis has not yet been tested in a cladistic framework.

The specimens described by de la Fuente et al. (2011) as *P. cerrobarcinae* are similar in size to the holotype of *P. argentinae* and smaller than the holotype of *Pr. portezuelae*. *Prochelidella cerrobarcinae* also exhibits posterior parts of the carapace that were not present in *P. argentinae* and *P. portezuelae*. Among some referenced specimens of *P. cerrobarcinae* (MPEF-PV- 3292–3294), the neural bone series is complete, nearly contiguous in dorsal view, and contiguous with the suprapygal in visceral view. This condition is present in some extant specimens of *Hydromedusa maximiliani* (see Wood and Moody 1976) but is not typical in chelid turtles, where the reduction of neural bones is the rule (Pritchard 1988).

4.1.3 *Mesoclemmys* Gray 1873

Type species: *Emys gibba* Schweigger 1812.

Species included: *M. dahli* (Zangerl and Medem 1958), *M. gibba* (Schweigger 1812), *M. heliostemma* McCord, Joseph-Ouni and Lamar 2001, *M. hogei* (Mertens 1967), *M. nasuta* (Schweigger 1812), *M. perplexa* Bour and Zaher (2005), *M. raniceps* (Gray 1855), *M. tuberculata* (Luederwaldt 1926), *M. vanderhaegei* (Bour 1973), and *M. zuliae* (Pritchard and Trebbau 1984).

Diagnosis: See McCord et al. (2001).

Mesoclemmys dahli (Zangerl and Medem 1958)

Holotype: Field Museum of Natural History, Chicago, USA FMNH 75980.

Diagnosis: See Zangerl and Medem (1958).

Distribution: Northeastern Colombia, in the vicinity of Sincelejo, Bolivar, Colombia (Zangerl and Medem 1958; Iverson 1992) (Fig. 4.4a).

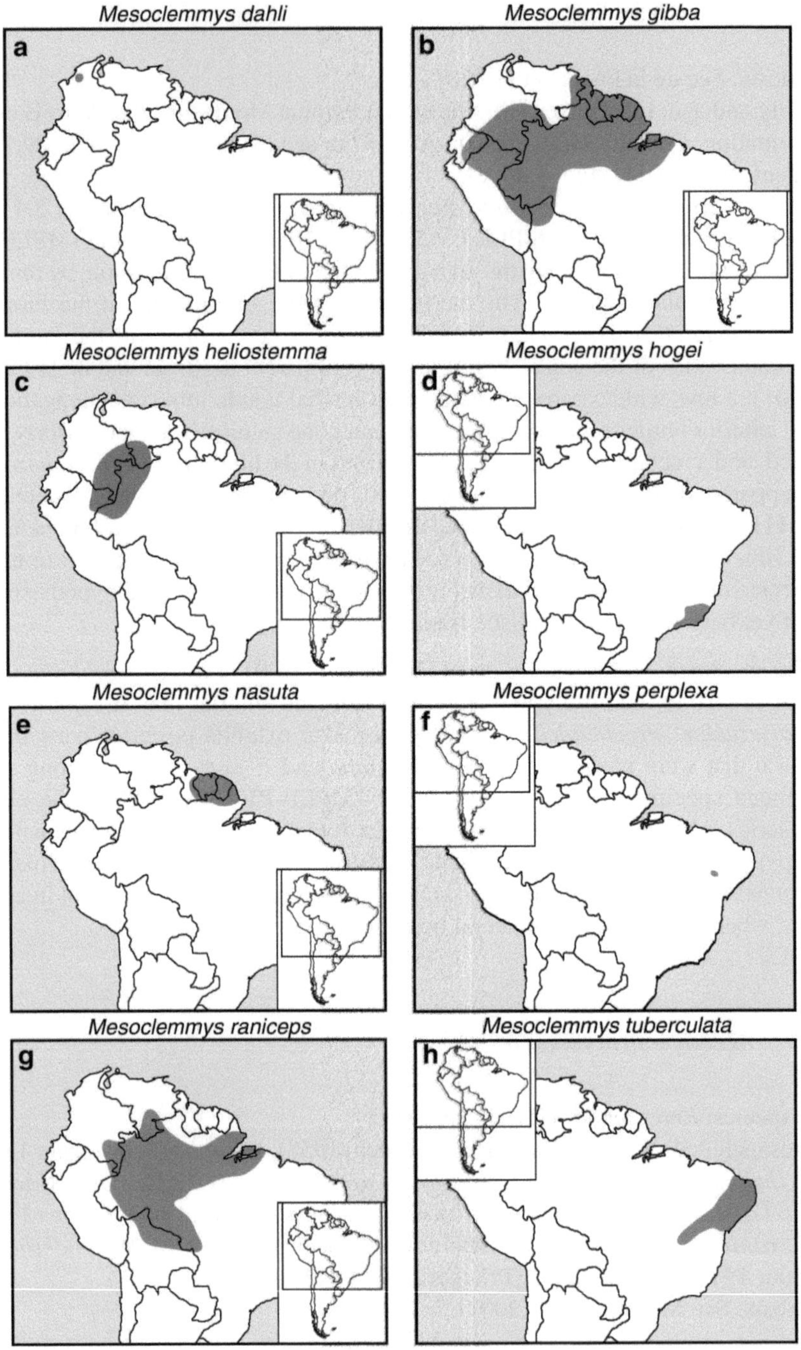

Fig. 4.4 South American distribution of *Mesoclemmys* (Chelidae) in recent times. (**a**) geographic distribution of *Mesoclemmys dahli*. (**b**) Geographic distribution of *Mesoclemmys gibba*. (**c**) Geographic distribution of *Mesoclemmys heliostemma*. (**d**) geographic distribution of *Mesoclemmys hogei*. (**e**) Geographic distribution of *Mesoclemmys nasuta*. (**f**) Geographic distribution of *Mesoclemmys perplexa*. (**g**) geographic distribution of *Mesoclemmys raniceps*. (**h**) Geographic distribution of *Mesoclemmys tuberculata*

Comments: This extant chelid species was erected by Zangerl and Medem (1958) on the basis of the holotype (an adult male) and three paratypes (a male, a juvenile male, and a hatchling). These authors presented a detailed description focused on external features and osteological characters of the skull (studied directly from the holotype), hyoid apparatus, shell, and other postcranial bones (studied from X-ray plates of the holotype). According to Zangerl and Medem (1958), *M. dahli* differs externally from *M. tuberculata* in that it lacks horny tubercles on the neck, has different streak patterns in the plastral coloration (uniform yellowish-whitish in *M. dahli* vs. yellow-brown with dark brown blotches in *M. tuberculata*), and has a narrow plastron relative to *M. tuberculata*. Other postcranial differences between the two species (e.g., different distal expansion of coracoids, the absence or presence of a well-defined inguinal buttress, different size opening between the two anteriormost costal ribs, and distinct size in metatarsals IV and V) are cited in Zangerl and Medem (1958). Zangerl and Medem's comparison of *M. dahli* and *M. nasuta* is restricted to external traits. These authors have demonstrated differences in coloration of the head, neck, carapace, and plastron and in limb robustness between the species. According to Vargas-Ramírez et al. (2012), *M. dahli* and *M. zuliae* are closely related species, with a fossil-calibrated molecular clock divergence time estimated of 10.6 million years.

Mesoclemmys gibba (Schweigger 1812)

Holotype: Muséum National d'Histoire Naturelle, Paris, France. MNHN 8756.
Diagnosis: See McCord et al. (2001).
Distribution: Orinoco to Amazon River basins in Colombia, eastern Ecuador, Peru, the Guianas, and northern Brazil (Iverson 1992) (Fig. 4.4b).
Comments: Originally the holotype of *Emys gibba* has been distinguished from the other toad-headed chelids, and it was referred to a different generic entity, *Mesoclemmys*. According to McCord et al. (2001), *Mesoclemmys gibba* shares some traits of both *Batrachemys* and *Phrynops*. More recently, Bour and Zaher (2005) referred the extant *Batrachemys* species to genus *Mesoclemmys*. As in *M. nasuta* and *M. dahli,* the snout of *M. gibba* is pointed, the parietal–squamosal arches are narrow, and the first neural bone is absent. However, *M. gibba* shares with *Phrynops* spp. parietals that are broadened above and parietal–squamosal arches that pass directly above but not posterior to the rear margin of the opisthotic bones. Although Bour and Zaher (2005) phylogenetic analysis using morphological characters did not resolve the relationships among the extant *Mesoclemmys* species, Vargas-Ramírez et al. (2012) cladogram, using molecular data, recovered *M. gibba* as the successive sister taxon of *M. dahli* and *M. zuliae*.

Mesoclemmys heliostemma (McCord, Joseph-Ouni and Lamar 2001)

Holotype: Smithsonian Institution, Washington, USA. USNM 541895.
Diagnosis: See McCord et al. (2001).
Distribution: Upper Amazon basin from southern Venezuela, western Brazil, northeastern Peru, and eastern Ecuador to southeastern Colombia (McCord et al. 2001) (Fig. 4.4c).

Comments: This species was recently named and described by McCord et al. (2001) based on a juvenile specimen (holotype) recovered from Pico de la Neblina locality (Venezuela–Brazil border) and specimens from the vicinity of Iquitos (Peru). McCord et al. (2001) provided a morphological description of this species and a detailed comparison between *M. heliostemma* and closely related species. Limited information concerning its habitat and ecology is also provided by McCord et al. (2001).

Mesoclemmys hogei (Mertens 1967)

Holotype: Naturmuseum und Forschung Institut Senckenberg, Frankfurt, Germany. SMF 62530.
Diagnosis: See McCord et al. (2001).
Distribution: Río Paraiba drainage of Rio de Janeiro and southern Minas Gerais, Brazil (Mittermeier et al. 1980) (Fig. 4.4d).
Comments: This extant species was named and described in detail by Mertens (1967) as *Phrynops hogei*. More recently, this species was transferred to the new genus *Ranacephala* by McCord et al. (2001). This taxonomic decision was made because McCord et al. (2001) and Rhodin in Reed et al. (1991) recognized that this species is osteologically and morphologically the most divergent taxon of *Phrynops*, particularly when it is compared with species of the *Phrynops geoffroanus* complex (see McCord et al. 2001) or *Rhinemys rufipes*. Recently, this species (*hogei*) has been referred to the genus *Mesoclemmys* by Bour and Zaher (2005) (see below).

Mesoclemmys nasuta (Schweigger 1812)

Holotype: Muséum National d'Histoire Naturelle, Paris, France. MHN 4140
Definition: See Bour and Pauler (1987).
Distribution: Amazon and Orinoco drainages of Colombia, Venezuela, Peru, Bolivia and Brazil, east through the Guianas (King and Burke 1989) (Fig. 4.4e).
Comments: According to Bour (1973), *M. nasuta* (Schweigger 1812) and *M. tuberculata* (Luederwaldt 1926) have been undistinguished for a long time. This is due to the absence of accurate definitions of these species. Bour (1973) believes that a variety of factors are involved in this misunderstanding. Firstly, Schweigger's (1812) holotype of *M. nasutus,* later described by Bour (1973), is a young specimen, and as noted by Bour (1973), young specimens differ greatly from adult specimens. Furthermore, there are probably several species similar in sympatry whose specimens are rare in the collections of important museums.

Mesoclemmys perplexa Bour and Zaher 2005

Holotype: Museu de Zoologia da Universidade de São Paulo, São Paulo, Brazil. MZUP 4111.
Diagnosis: See Bour and Zaher (2005).
Distribution: Restricted to southeast of the State of Piauí, northeastern Brazil (Bour and Zaher 2005) (Fig. 4.4f).

Comments: A detailed morphological description based on external characters of *Mesoclemmys perplexa* was recently published by Bour and Zaher (2005). Although the strict consensus of the cladistic analysis performed by the authors supports a basal position of *Phrynops geoffroanus* and recovers *Rhinemys rufipes* as the sister taxon of a clade that includes the remaining toad-headed chelid turtles, this parsimony analysis disagrees with the resolution of the cladogram presented by McCord et al. (2001). Contrarily to McCord et al.'s (2001) phylogenetic hypothesis, the topology that resulted from Bour and Zaher's (2005) phylogeny recovers a polytomy that includes *M. perplexa* and the remaining toad-head species of McCord et al. (2001): *Mesoclemmys gibba*, *Bufocephala vanderhagei*, *Ranacephala hogei*, *Batrachemys dalhi*, *Ba. heliostemma*, *Ba. nasuta*, *Ba. raniceps*, *Ba. tuberculata*, and *Ba. zuliae*. As a result of Bour and Zaher's (2005) parsimony analysis, the taxonomic scheme suggested by McCord et al. (2001) for the toad-headed chelid (five new genera) is not accepted by the above-mentioned authors, who proposed to assign the toad-headed species to the genus *Mesoclemmys*.

Mesoclemmys raniceps (Gray 1855)

Lectotype: British Museum of Natural History London, United Kingdom. The specimen BMNH 1947.3.5.92 is designed lectotype of *M. raniceps* by Bour and Pauler 1987 because the holotype is not stated.

Definition: See Bour and Pauler (1987).

Distribution: Upper Orinoco to Amazon River basins in eastern Colombia, southern Venezuela, Peru, Brazil, and Bolivia (Iverson 1992) (Fig. 4.4g).

Comments: The Amazon toad-headed turtle was first named *Hydraspis raniceps* by Gray (1855) and referred to *nasuta* by Boulenger (1989). Bour and Pauler (1987) resurrected *raniceps* as a full species of the genus *Phrynops*. More recently *raniceps* was transferred to genus *Batrachemys* (McCord et al. 2001) or to genus *Mesoclemmys* (Bour and Zaher 2005). According to Bour and Pauler (1987) *Phrynops wermuthi* Mertens 1969 is included in the synonymy of *M. raniceps*. A brief summary on their biology is reported by Medem (1960).

Mesoclemmys tuberculata (Luederwaldt 1926)

Syntypes: Museu de Zoologia Universidade de São Paulo, São Paulo, Brazil. MZUSP 43 and 81. Bour and Pauler (1987) designed MZUSP 43 the lectotype, because MZUSP 81 is lost.

Definition: See Bour and Pauler (1987).

Distribution: Eastern Brazil, Río San Francisco and adjacent basins (Iverson 1992) (Fig. 4.4h).

Comments: This toad-head, side-necked turtle was first described as *Rhinemys tuberculata* by Luederwaldt (1926) and later described as *Phrynops tuberculata* by Bour (1973) and Bour and Pauler (1987), who provided additional information about its taxonomy and biology. McCord et al. (2001) transferred this species to genus *Batrachemys*. More recently, Bour and Zaher (2005) referred *tuberculata* to genus *Mesoclemmys*.

Mesoclemmys vanderhaegei (Bour 1973)

Holotype: Muséum National d'Histoire Naturelle, Paris, France. MNHN 1977–50.
Diagnosis: See McCord et al. (2001).
Distribution: Río Paraguay and Río Parana drainages of Paraguay, Brazil, and probably northern Argentina and Bolivia (Bour and Pauler 1987) (Fig. 4.5a).
Comments: This species was initially named as a subspecies of *Phrynops tuberculatus* (= *Phrynops tuberculatus vanderhaegei*) by Bour (1973). This author presents a differential diagnosis of this taxonomic entity and notes the affinities with *B. vanderhaegei*. More recently, Bour and Pauler (1987) referred this taxon as a species of the genus *Phrynops* (= *Phrynops vanderhaegei*). In that paper the authors described the specific characters of this species and supplement these morphological data with taxonomical, biological, and distributional information. Finally, McCord et al. (2001) suggested assigning this species to a new genus (*Bufocephala*). Contrarily, Bour and Zaher (2005) did not agree with this taxonomic decision and referred *vanderhagei* to genus *Mesoclemmys*.

Mesoclemmys zuliae (Pritchard and Trebbau 1984)

Holotype: Florida Museum of Natural History, University of Florida, Gainesville, USA UF 53439.
Diagnosis: See Pritchard and Trebbau (1984).
Distribution: Drainage of the Ciénaga de Juan Manuel de Aguas Claras, downstream into the western side of Maracaibo Lake, Venezuela (Pritchard and Trebbau 1984; King and Burke 1989) (Fig. 4.5b).
Comments: As noted by Pritchard and Trebbau (1984), the head of *M. zuliae* differs from that of *M. nasuta* in size, proportion, and outline. Pritchard and Trebbau (1984) compared skulls of two adult specimens (male and female) of *M. zuliae* to *M. nasutus* and found important morphological differences. Relative to *M. nasutus*, they found the following characteristics in *M. zuliae*: greater skull elongation, wider postorbital arcades, stronger cusp on the front of the upper rhamphotheca, a strong anterior process of the basisphenoid that almost completely separated the pterygoids, vomer well separated from premaxillae by the maxillae, foramen magnum much higher than wide, foramen nervi trigemini rather broadly oval, and a unique hemispheral boss the inner edge of the upper triturating surface. A detailed anatomical description of head, extremities, and shell coloration, as well as relevant biological data of this species, is included in Pritchard and Trebbau (1984). According to Vargas-Ramírez et al. (2012), *M. zuliae* and *M. dahli* are closely related species, and their estimated divergence time is correlated with the uplift of the Serranía de Perijá in the Late Miocene and Pliocene.

4.1.4 *Phrynops* Wagler 1830

Type species: *Emys geoffroana* Schweigger 1812.
Species included: *Phrynops geoffroanus* (Schweigger 1812), *P. hilarii* (Duméril and Bibron 1835), and *P. williamsi* Rhodin and Mittermeier 1983.

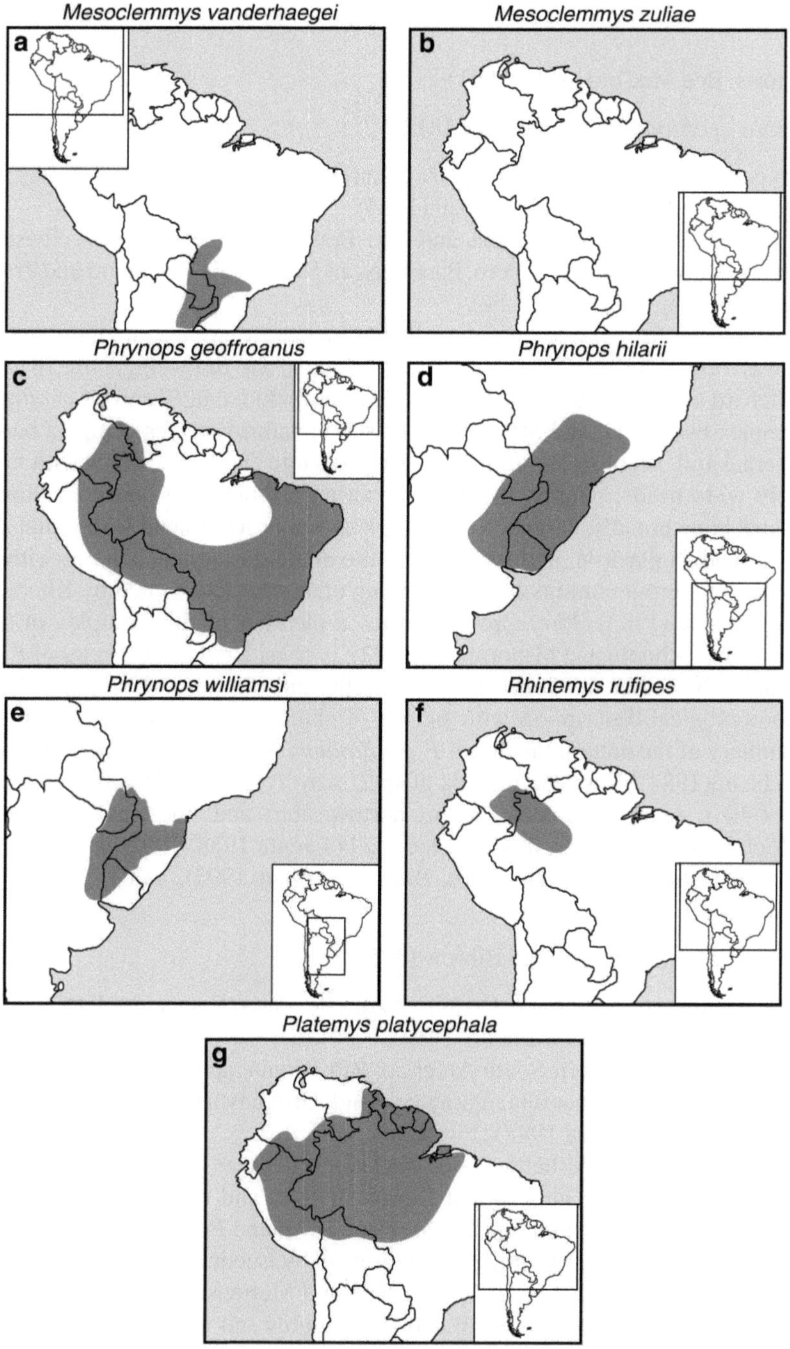

Fig. 4.5 South American distribution of Chelidae in recent times. (**a**) geographic distribution of *Mesoclemmys vanderhaegei*. (**b**) Geographic distribution of *Mesoclemmys zuliae*. (**c**) Geographic distribution of *Phrynops geoffroanus*. (**d**) Geographic distribution of *Phrynops hilarii*. (**e**) Geographic distribution of *Phrynops williamsi*. (**f**) Geographic distribution of *Rhinemys rufipes*. (**g**) Geographic distribution of *Platemys platycephala*

Diagnosis: See McCord et al. (2001).

Phrynops geoffroanus (Schweigger 1812)

Holotype: Museum National d'Histoire Naturelle, Paris, France. MNHN 9417.
Diagnosis: See Pritchard and Trebbau (1984).
Distribution: Orinoco to Amazon and São Francisco to Paraná River basins of Brazil, Bolivia, Colombia, Peru, Paraguay, and Venezuela (Pritchard and Trebbau 1984; Iverson 1992) (Fig. 4.5c).
Comments: Geoffroy's side-necked turtle has had an extensive synonymy list since it was originally described as *Emys geoffroana* by Schweigger in 1812 (see Pritchard and Trebbau 1984). This species was included in *Phrynops geoffroanus* complex by Rhodin and Mittermeier (1983), a natural species complex based in external and osteological similarities such as large and broad shells with moderately wide heads, neurals usually numbering six or seven and contacting the nuchal bone broadly, carapace either dark brown or reticulated with radial markings, reddish plastron, and yellow or white and either immaculate or with scattered dark vermiculations or spots, among other characters cited by Rhodin and Mittermeier (1983). *Phrynops tuberosus,* a member of this complex in accordance with Rhodin and Mittermeier (1983), is considered a subspecies of *P. geoffroanus* by other authors (i.e., Wermuth and Mertens 1961). A detailed morphological description with figures of skull, carapace, and plastron and a summary of the natural history of *P. geoffroanus* was published by Pritchard and Trebbau (1984). Neogene records of specimens referred to *P. geoffroanus* complex have been documented from northwestern and northeastern Argentina (Wieland 1923; Gasparini et al. 1986; de la Fuente 1988, 1992; Broin and de la Fuente 1993; Cione et al. 2000), Brazil (Kischlat 1993), and Uruguay (Perea et al. 1996).

Phrynops hilarii (Duméril and Bibron 1835)

Holotype: Museum National d'Histoire Naturelle, Paris, France. MNHN 8757.
Definition: See Ernst and Barbour (1989).
Distribution: Southeastern South America, Río Paraná, and Paraguay drainages of southern Brazil, Argentina, Paraguay, and Uruguay (King and Burke 1989; Iverson 1992; Cabrera 1998) (Fig. 4.5d).
Comments: The spotted-bellied side-necked turtle was originally described as *Platemys hilarii* by Duméril and Bibron in 1835 and referred to as *Hydraspis hilarii* Gray 1844, *Spatulemys lasalae* Gray 1872, and *Phrynops hilarii* Stejneger 1909 or as a variety of *Hydraspis geoffroyana* by Luederwaldt 1926 or a subspecies of *Phrynops geoffroanus* by Wermuth and Mertens (1961). A detailed morphological description and skull and shell figures and a summary of its natural history are presented by Cabrera (1998).

Phrynops williamsi Rhodin and Mittermeier 1983

Holotype: Museum of Comparative Zoology, Harvard University, Massachusetts, USA. MCZ 64135.

Diagnosis: See Rhodin and Mittermeier (1983).

Distribution: Low-lying areas of eastern coastal of Santa Catarina and Rio Grande do Sul in Brazil, northern half of Uruguay, Misiones, Corrientes, and Entre Ríos provinces in Argentina (Rhodin and Mittermeier 1983; Richard and de la Fuente 1992) (Fig. 4.5e).

Comments: This species was named by Rhodin and Mittermeier (1983). In this paper the authors described in detail the external morphology and skull osteology and made a brief characterization of the cervical formulae and carapace neural bone configuration of *P. williamsi*. In accordance with Rhodin and Mittermeier's paper, the characteristics of this species as discussed by these authors distinguish it from other members of the *Phrynops geoffroanus* complex (sensu Rhodin and Mittermeier 1983) (= *Phrynops* sensu stricto after McCord et al. 2001). Although it is slightly different in skull anatomy, *P. williamsi* is hardly distinguishable on external morphology from *P. geoffroanus geoffroanus* and *P. geoffroanus tuberosus*. Fortunately to herpetologist and collectors, *P. williamsi* though sympatric with *P. hilarii* is allopatric with respect to *P. geoffroanus* (see Rhodin and Mittermeier 1983, Fig. 9).

4.1.5 *Rhinemys* Wagler 1830

Type species: *Emys rufipes* Spix 1824.

Diagnosis: See McCord et al. (2001).

Species included: Only *Rhinemys rufipes* (Spix 1824).

Rhinemys rufipes (Spix 1824)

Holotype: Zoologisches Sammlungdes Bayerischen Staates, Münich, Germany. ZZM 3006.

Diagnosis: See McCord et al. (2001).

Distribution: Mid-upper Amazon Basin in Brazil, Colombia, and Peru (Lamar and Medem 1984) (Fig. 4.5f).

Comments: This species was originally named *Emys rufipes* by Spix (1824) and later transferred to *Phrynops* by Gray (1844) and to *Hydraspis* by Boulenger (1989). Most recent researchers follow Gray (1844) generic assignation (Wermuth and Mertens 1961; Medem 1973; Pritchard 1979; King and Burke 1989). More recently, McCord et al. (2001) resurrected the genus name *Rhinemys* Wagler 1830 for this species. These researchers justified the removal of the species *rufipes* from genus *Phrynops* because several characters differentiate this species from other members of the complex. Phylogenetic relationships based on serological data from *Rh. rufipes* were proposed by Frair (1982), and more recently, it was reviewed by Lamar and Medem (1984). In this latter paper, the authors reported additional data on dimensions, external morphology, coloration, maximum size, sexual dimorphism, ecology, behavior in captivity, and distribution within Colombia.

4.1.6 *Platemys* Wagler 1830

Type species: *Testudo platycephala* Schneider 1792.
Diagnosis: Pritchard and Trebbau (1984).
Species included: Only *Platemys platycephala.*

Platemys platycephala (Schneider 1792)

Holotype: Undesigned.
Definition: See Ernst (1987) and Ernst and Barbour (1989) and morphological description in Pritchard and Trebbau (1984).
Distribution: This species range is the Caribbean drainages of eastern Venezuela and the Guianas, and the Amazon Basin of Bolivia, Brazil, Colombia, Ecuador, and Peru (Ernst 1987) (Fig. 4.5g).
Comments: The binominal assignation of this species was initially unclear since Schneider (1792), in his original description, used both *Testudo platycephala* and *T. planiceps*. Pritchard and Trebbau (1984) argue that the International Committee of Zoological Nomenclature should accept the name *platycephala* rather than the less common *planiceps*. Ernst (1987) summarized the edited data for this species. Additional information regarding the ecology, habitat, reproduction, feeding, and conservation of this species is summarized by Pritchard and Trebbau (1984). Two subspecies are recognized: *Platemys platycephala platycephala* (Schneider 1972) and *Platemys platycephala melannonota* (Ernst 1984) (see Ernst 1984, 1987, and references therein).

4.1.7 *Salamanchelys* Bona 2006

Type species: *Salamanchelys palaeocenica* Bona 2006.
Diagnosis: See Bona 2006.
Species included: *Salamanchelys palaeocenica* Bona 2006.

Salamanchelys palaeocenica Bona 2006

Holotype: Museo Paleontológico Egidio Feruglio, Trelew, Argentina. MPEF 598, complete plastron articulated with fragmentary carapace (posterior portion of the carapace and left margin missing).
Diagnosis: See Bona (2006).
Locality, horizon, and age: Cerro Hansen, Chubut Province, Argentina. According to Bona et al. (1998), the turtle remains came from a Danian level underlying the Banco Negro Inferior included in the Salamanca Formation (Lesta and Ferello 1972) (Fig. 4.6a, b).
Comments: This is a small chelid turtle named and described by Bona (2006), based on the holotype and several isolated shell bones as referenced material in Bona (2006). Previously, Staesche (1929) referred a plastral fragment and a first right

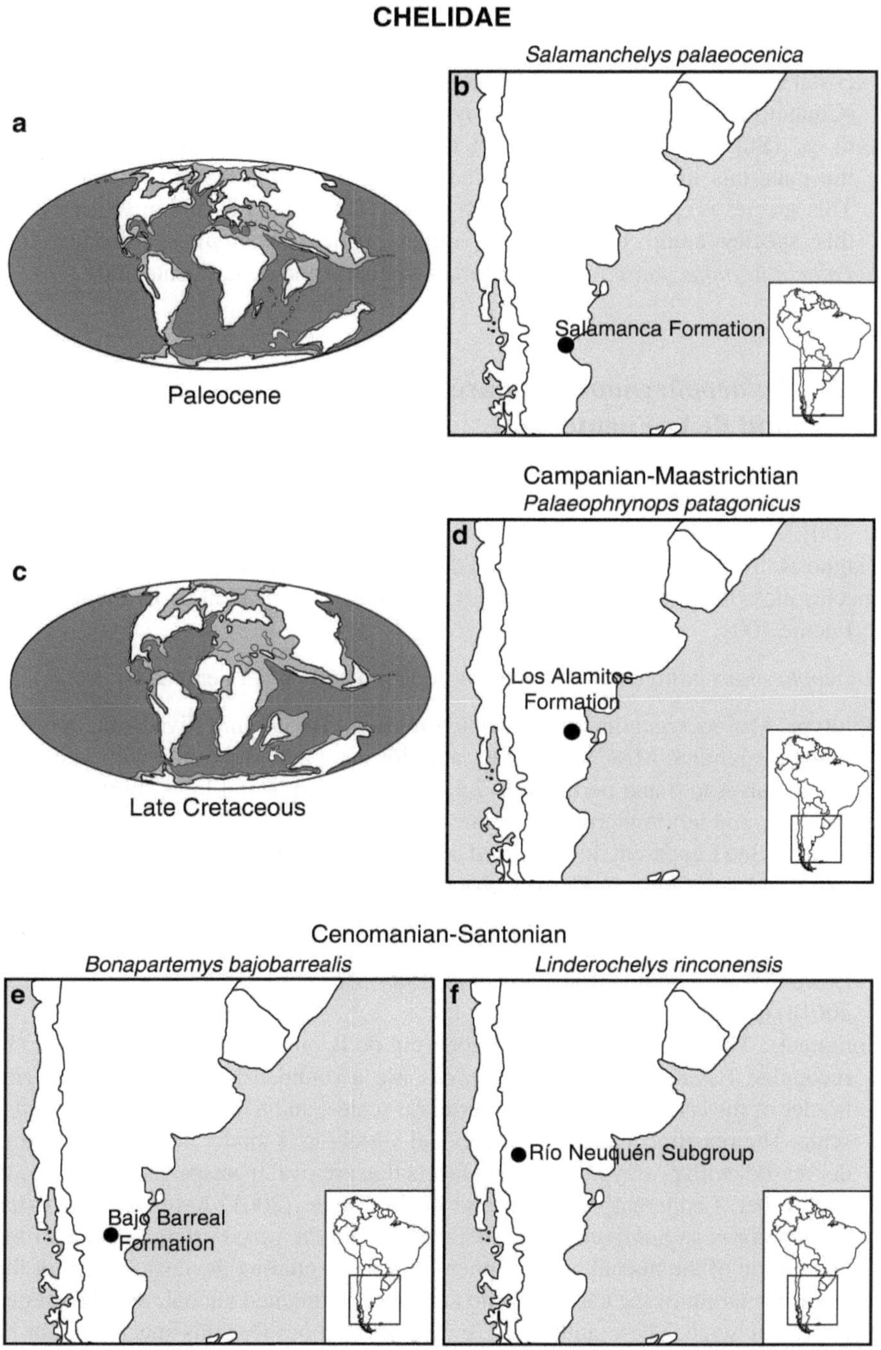

Fig. 4.6 Geographic and stratigraphic distribution of Chelidae during the Palaeocene and Late Cretaceous. (**a**) Paleogeographic reconstruction of continents during the Palaeocene. (**b**) Geographic and stratigraphic location of *Salamanchelys palaeocenica*. (**c**) Paleogeographic reconstruction of continents during the Late Cretaceous. (**d**) Geographic and stratigraphic location of *Paleophrynops patagonicus*. (**e**) Geographic and stratigraphic location of *Bonapartemys bajobarrealis*. (**f**) Geographic and stratigraphic location of *Linderochelys rinconensis*

costal bone fragment recovered from the Salamanca Formation at Punta Peligro (Chubut Province, Argentina) to Emydidae as "Novum Genus, cf. *Gyremys* Hay n. sp." (Staesche 1929: 116–118; pl 18, Figs. 8–11). According to Bona (2006), the materials illustrated by Staesche belong to *Salamanchelys palaeocenica*. This species is referred to Chelidae by Bona (2006) and this author suggests that this species could be related to other extinct *Phrynops* s.l. species like *Palaeophrynops patagonicus* Lapparent de Broin and de la Fuente 2001.

4.1.8 *Palaeophrynops* Lapparent de Broin and de la Fuente 2001

Type species: *Palaeophrynops patagonicus* Lapparent de Broin and de la Fuente 2001.
Diagnosis: See Lapparent de Broin and de la Fuente (2001) and de la Fuente (2007).
Species included: Only *Palaeophrynops patagonicus* Lapparent de Broin and de la Fuente 2001.

Palaeophrynops patagonicus Lapparent de Broin and de la Fuente 2001

Holotype: Museo Argentino de Ciencias Naturales "Bernardino Rivadavia," Buenos Aires, Argentina. MACN–RN 906, anterior carapace, right side with lateral of the costals 4 to 6 and peripherals 7 and 8, anterior plastral lobe, fourth cervical vertebra, and left humerus and femur.
Diagnosis: See Lapparent de Broin and de la Fuente (2001) and de la Fuente (2007).
Locality and age: Western Cerro Colorado, northwestern corner of "Estancia Los Alamitos," close to Arroyo Verde, Valcheta Department, Río Negro, Argentina. Middle part of the Los Alamitos Formation, Upper Cretaceous (Upper Campanian–Lower Maastrichtian) (Bonaparte et al. 1984; Bonaparte 1990; Hugo and Leanza 2001b) (Fig. 4.6c, d).
Comments: The characters that led Lapparent de Broin and de la Fuente (2001) to recognize *Palaeophrynops patagonicus* are a rounded and elongated anterior border of the carapace, narrowed vertebral scales, and a larger, dilated intergular scale. The opisthocoelous fourth cervical vertebrae is moderately elongated and depressed, with postzygapophyseal facets that are oval in shape and very close to each other. Lapparent de Broin and de la Fuente (2001) have suggested that *Palaeophrynops patagonicus* is a member of the *Phrynops* group because of the elongation of the nuchal and peripheral bones (excluding the costal ones) of the anterior margin of the carapace; the narrowed, elongated nuchal; and the dilated intergular scale. These authors have also pointed out that this species might be related to the large extant *Phrynops* species of the *Phrynops geoffroanus* complex (Rhodin and Mittermeier 1983), included in the subgenus *Phrynops* (Pritchard and Trebbau 1984) or in the genus *Phrynops* sensu stricto (McCord et al. 2001).

4.1.9 *Bonapartemys* Lapparent de Broin and de la Fuente 2001

Type species: *Bonapartemys bajobarrealis* Lapparent de Broin and de la Fuente 2001.
Diagnosis: See Lapparent de Broin and de la Fuente (2001) and de la Fuente (2007).
Species included: *Bonapartemys bajobarrealis* Lapparent de Broin and de la Fuente 2001.

Bonapartemys bajobarrealis Lapparent de Broin and de la Fuente 2001

Holotype: Museo Argentino de Ciencias Naturales "Bernardino Rivadavia," Buenos Aires, Argentina. MACN-CH-1469, carapace, plastron, remains of girdles, limbs, and tail.
Diagnosis: See Lapparent de Broin and de la Fuente (2001) and de la Fuente (2007).
Locality and age: Located 6 km north of "Estancia Ocho Hermanos," Sierra de San Bernardo, Sarmiento Department, Chubut Province, Argentina. Upper part of the Lower Member of the Bajo Barreal Formation; Upper Cretaceous, Turonian–Campanian (Martínez et al. 1986; Barcat et al. 1989; Leanza 1999), or Cenomanian–Turonian (Bridge et al. 2000) (Fig. 4.6c, e).
Comments: The holotype of this species was recovered by Dr. José Bonaparte from the lower member of the Bajo Barreal Formation (see section in Martínez et al. 1986) at "Estancia Ocho Hermanos" locality (see above).[3] Subsequently, Lapparent de Broin and de la Fuente (2001) formally designated new species *Bonapartemys bajobarrealis* on the basis of the specimen MACN-CH-1469. These authors assigned this species to the clade Chelidae on account of the suture connection between the pelvic girdle and the shell (pleurodiran condition, see Gaffney and Meylan 1988 and references therein), associated with narrowed vertebral scales 2 through 4, and the loose connection between costal and peripheral bones (see Lapparent de Broin and de la Fuente 2001). Lapparent de Broin and de la Fuente (2001) noted that *B. bajobarrealis* is similar to the large extant *Phrynops* (*Phrynops* sensu strict after McCord et al. 2001), with neural bones like those of the *Phrynops geoffroanus* complex (see Rhodin and Mittermeier 1983), but with other plesiomorphic characters including a wide nuchal bone, slightly elongated anterior peripheral bones with axillary process at peripheral 2–3, absence of a dilated intergular scale, and the presence of mesoplastra (see de la Fuente 2007). However, a different systematic assignment to *Bonapartemys bajobarrealis* is proposed by Cadena and Bloch (2009), whose abstract introduced a new phylogenetic hypothesis regarding the relationships of panpleurodires and select stem Testudines and suggested that *B. bajobarrealis* is more closely related to Panpelomedusoides than chelids.

[3]Courtesy of Indiana University Press.

4.1.10 *Linderochelys* de la Fuente, Calvo and González Riga 2007

Type species: *Linderochelys rinconensis* de la Fuente, Calvo and González Riga 2007.
Diagnosis: See de la Fuente, Calvo and González Riga (2007).
Species included: Only *Linderochelys rinconensis* de la Fuente, Calvo and González Riga 2007.

Linderochelys rinconensis de la Fuente, Calvo and González Riga 2007

Holotype: Museo Argentino Urquiza de Rincón de Los Sauces, Neuquén, Argentina. MAU-PV-LL-69, a fragment of the suprapygal and the eighth right costal bone of the carapace, a partially preserved plastron, one cotyle and one condyle (probably belonging to a second through fourth opisthocoelous vertebrae or a seventh biconcave), as well as fragments of two postzygapophyses of cervical vertebrae, left humerus, radius and proximal carpal elements and right metacarpals and digits, the pelvic girdle, femora and fibula, and right tibia.
Diagnosis: See de la Fuente, Calvo and González Riga (2007).
Locality and age: The specimen was recovered in mudstones and fine sandstones that crop out at Loma del Lindero site, 10 km west of Rincón de los Sauces City, Neuquén Province, Patagonia, Argentina. These rocks belong to the Río Neuquén Subgroup, Neuquén Group, Late Cretaceous, and Late Turonian–Coniacian according to Leanza (1999), Leanza and Hugo (2001), and Leanza et al. (2004) (Fig. 4.6c, f).
Comments: Although unnamed species, represented by isolated cervical vertebrae, which show similar plesiomorphic characters present in the Australasian extant species of the *Emydura–Elseya* group have been recognized from the Allen Formation (Upper Cretaceous) of Patagonia (Broin and de la Fuente 1993, pl. 1, Figs. 2–3), all the recognized chelid taxa recovered from Early and Upper Cretaceous horizons of Patagonia are considered closely related to the South American short- and long-necked chelids (Lapparent and de la Fuente 2001; de la Fuente et al. 2001, 2007, 2011). On account of the morphology of its preserved cervical vertebrae, *Linderochelys rinconensis* was tentatively assigned to the *Phrynops* group (sensu Broin and de la Fuente 1993) by de la Fuente et al. (2007). *Linderochelys rinconensis*' cervical vertebrae are similar to those of the extant species of the *Phrynops* group. These vertebrae are slightly elongated and lower than the cervical vertebrae of the Australasian short-necked species and are characterized by the presence of ventral keels that are curved and reduced posteriorly relative to the central length. *Linderochelys rinconensis* is a short-necked chelid characterized by an elongated abdominal seam. It clearly differs from the South American extant *Phrynops* (sensu lato) in the presence of cuneiform mesoplastra crossed by pectoral–abdominal sulcus. Likewise, *Li. rinconensis* also differs from the Upper Cretaceous chelids hitherto assigned to the *Phrynops* group based on several characters (see diagnosis in de la Fuente et al. 2007 and Lapparent de Broin and de la Fuente 2001).

Broin and de la Fuente (1993) have used the *Phrynops* group (= *Phrynops* sensu lato) to refer to a South American group of short-necked chelids. Among the extant forms, this group includes (a) *Phrynops* s.s. with three poorly defined extant subgenera

(*Phrynops* (Wagler), *Mesoclemmys* (Gray), and *Batrachemys* (Stejenger)) sensu Bour (1973) and Bour and Pauler (1987); (b) *Acanthochelys*, excluded from *Platemys auctoris*; (c) *Platemys* (Wagler) restricted to *Pl. platycephala* (Schneider). According to Broin and de la Fuente (1993), *Phrynops* (s.l.) is characterized by two synapomorphies: the dorsal inflation of the antrum squamosum and the inclination of the columella–fenestra ovalis axis. The species included in *Phrynops* (s.l.) are characterized by a short neck as is true of members of the *Emydura–Elseya* group but also by cervical vertebrae that are slightly more elongated and short dorsoventrally than Australasian short-necked species and with ventral keels that are curved and reduced posteriorly in relation to the central length. According to de la Fuente et al. (2007), the cotyle fragment preserved in the holotype of *Linderochelys rinconensis* is similar in outline to the cotyle of some extant species of *Phrynops* (s.s.) (i.e., *Phrynops geoffroanus tuberosus*).

Broin and de la Fuente's concept of *Phrynops* (sensu stricto) is not comparable to *Phrynops* (sensu stricto) of McCord et al. (2001), the last reviewers of *Phrynops*. *Phrynops* sensu (McCord et al. 2001) is restricted to *Phrynops geoffroanus* complex of Rhodin et al. (1983). Prior to McCord's work, several authors explored the relationships among the extant species of *Phrynops* (s.l.). Gaffney (1977) was unable to find derived cranial characters among some species of *Phrynops* available for his study [*Phrynops* (*P.*) *geoffroanus* (Schweigger), *P.* (*M.*) *gibbus* (Schweigger), and *P.* (*B.*) *nasutus* (Schweigger)]. The monophyly of the genus *Phrynops* (s.l.) seemed to be supported by genetic data published by Frair (1980, 1982) and Reed et al. (1991). Likewise, Shaffer et al. (1997) provided a phylogeny where the genus *Phrynops* (s.l.) was found to be a monophyletic taxon. This conclusion was reached using mitochondrial DNA sequencing, as well as morphological and paleontological data. Contrary to this proposal, Seddon et al. (1997) strongly supported paraphyly among the subgenera of *Phrynops* (s.l.). This conclusion was sustained after Seddon et al. (1997) sequenced 411 mitochondrial 12 S rRNA nucleotides of 16 species assigned to 11 chelid genera. Analysis using parsimony and neighbor-joining algorithms supported generic differentiation among the subgenera of *Phrynops* (s.l.). In this context, *Mesoclemmys* was proposed as the sister taxon of *Batrachemys* and *Phrynops* as closely related to the genus *Chelus*. Likewise, Georges et al. (1998) molecular data supported Seddon et al.'s (1997) conclusion regarding the paraphyly of the genus *Phrynops* (s.l.) and elevated the subgenera *Phrynops*, *Mesoclemmys,* and *Batrachemys* once again to genera. Finally, McCord et al. (2001) demonstrated the full recognition of these taxa as genera based on morphometric analysis, skull osteology, and nuclear and mitochondrial gene sequences. These three distinct taxa were previously demoted to subgenera and then synonymized with *Phrynops* (Pritchard 1979; Pritchard and Trebbau 1984).

4.1.11 *Chelus* Duméril 1806

Type species: *Testudo fimbriata* Schneider 1783.

Species included: *Chelus fimbriata* (Schneider 1783), *Chelus colombianus* Wood 1976, and *Chelus lewisi* Wood 1976.

Diagnosis: See Gaffney (1977) and Pritchard and Trebbau (1984).

Chelus fimbriata (Schneider 1783)

Holotype: Not traced.
Diagnosis: See Pritchard and Trebbau (1984).
Distribution: Northern and central South America; the Amazon and Orinoco drainages of Brazil, Ecuador, Peru, Bolivia, Colombia, Venezuela, French Guiana; and the middle and upper Essequibo system of Guiana (Pritchard and Trebbau 1984; Iverson 1992; Pritchard 2008) (Fig. 4.7a, b).
Comments: The old taxonomic history of the "matamata" turtle was detailed by Pritchard (2008). According to the ICZN (International Commission of Zoological Nomenclature 1963), the scientific name of this bizarre chelid species is *Testudo fimbriata* Schneider 1738. This species was assigned to the new genus *Chelus* by Duméril (1806) and fully described by Pritchard and Trebbau (1984) who provided skull and shell figures and by Pritchard (2008) and Gaffney (1977) who provided skull figures. Although no subspecies of *Chelus fimbriata* have been recognized, the above-mentioned authors and Sánchez Villagra et al. (1995) recognized two different morphotypes when they made comparisons between matamatas from the Amazon and Orinoco drainages (e.g., parallel-sided or slightly constricted carapace vs. ovoid). However, in accordance with Sánchez Villagra et al. (1995), other characters do not show a similar distribution (e.g., the intergular failing to separate the gulars). Updated information regarding geographic variation, habitat, ecology, reproductive biology, population status, and conservation is summarized by Pritchard and Trebbau (1984) and Pritchard (2008).

Chelus colombianus Wood 1976

Holotype: University of California Museum of Paleontology, California, USA. UCMP 78762, a nearly complete shell.
Hypodigm: Additional specimens belonging to UCMP and to GMB (Museum of Geological Survey of Colombia) (see Wood 1976).
Diagnosis: See Wood (1976).
Locality and age: According to Wood (1976), the specimens included in the hypodigm of *C. colombianus* were found in the vicinity of three localities: Coyaima, Carmen de Apicalá, and Villavieja (upper Magdalena river Valley, Colombia) from Miocene sediments of the Villavieja Formation (Fig. 4.7c, d).
Comments: *Chelus colombianus* was named by Wood (1976) on the basis of several complete and incomplete shells and possesses characters that differentiate it from *C. fimbriata*. As it was pointed out by Wood (1976), among these differential traits, the recessed intergular scale (withdrawn from the anterior plastron margin) is a chelonian character only present in species of the related Australasian genus *Chelodina*. According to Wood (1976), the intergular configuration in *C. colombianus* is either hexagonal, as is typical in extant *Chelodina* species, or octagonal in two specimens where anterior plastral scales are available.

Chelus lewisi Wood 1976

Holotype: Museo de Ciencias Naturales de Caracas, Caracas, Venezuela. MCNC 239, a complete shell.

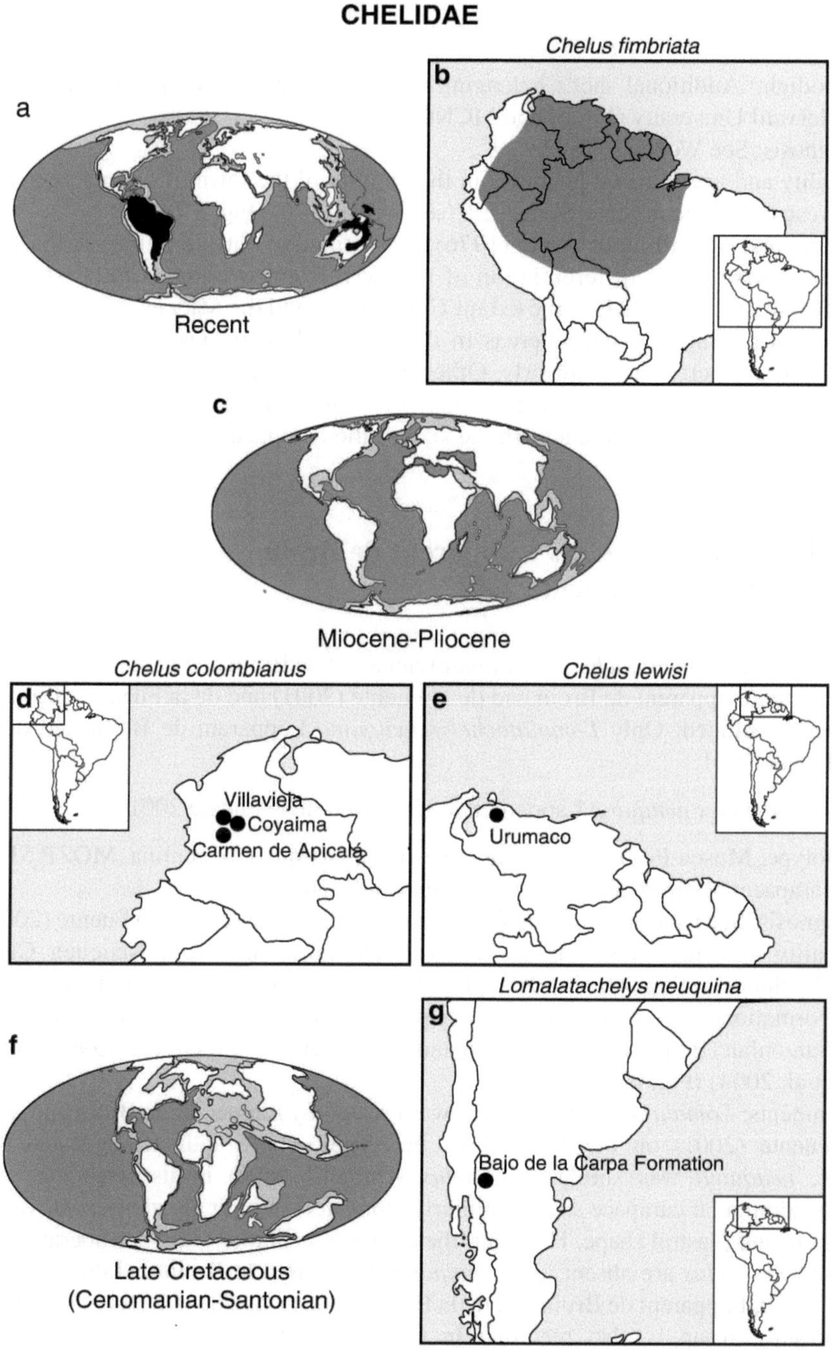

Fig. 4.7 Geographic and stratigraphic distribution of *Chelus* (Chelidae) and related taxa. (**a**) global distribution of Chelidae. (**b**) Geographic distribution of *Chelus fimbriata* in South America. (**c**) Paleogeographic reconstruction of continents during the Miocene–Pliocene. (**d**) Geographic and stratigraphic location of *Chelus colombianus*. (**e**) Geographic and stratigraphic location of *Chelus lewisi*. (**f**) Paleogeographic reconstruction of continents during the Late Cretaceous. (**g**) Geographic and stratigraphic location of *Lomalatachelys neuquina*

Hypodigm: Additional shells belonging to the Museum of Comparative Zoology, Harvard University (MCZ) and MCNC.
Diagnosis: See Wood (1976).
Locality and age: Several localities in the vicinity of the town of Urumaco, northwestern of Falcon State, Venezuela (see Wood 1976) (Fig. 4.7c, e).
Comments: According to Wood (1976) the distinctive outline of the carapace in *Ch. lewisi* allows differentiation of *C. lewisi* from two other *Chelus* species. The shape of the shell of the extant *C. fimbriata* and the Miocene *C. colombianus* are parallel sided, whereas in the Venezuelan species, the width of the carapace increases posteriorly. Other distinctive characters that may serve to distinguish *C. lewisi* from the extant *Ch. fimbriata* are the shape of the fist neural bone and differences in the size of the carapace.

4.1.12 *Lomalatachelys* Lapparent de Broin and de la Fuente 2001

Type species: *Lomalatachelys neuquina* Lapparent de Broin and de la Fuente 2001.
Diagnosis: Lapparent de Broin and de la Fuente (2001) and de la Fuente (2007).
Species included: Only *Lomalatachelys neuquina* Lapparent de Broin and de la Fuente 2001.

Lomalatachelys neuquina Lapparent de Broin and de la Fuente 2001

Holotype: Museo Profesor Olsacher de Zapala, Neuquén, Argentina. MOZP 5117, carapace, plastron, and remains of pubis and ischia.
Diagnosis: See Lapparent de Broin and de la Fuente (2001) and de la Fuente (2007).
Locality and age: Loma de la Lata (75 km northwest from Neuquén City), Confluencia Department, Neuquén Province, Argentina. Bajo de la Carpa Formation, Río Colorado Subgroup, Neuquén Group, Upper Cretaceous, Santonian (Bonaparte 1991; Broin and de la Fuente 1993; Leanza 1999; Leanza et al. 2004) (Fig. 4.7f, g).
Comments: *Lomalatachelys neuquina* was created by Lapparent de Broin and de la Fuente (2001) on the basis of its holotype.[4] These authors suggested that *L. neuquina* was similar to *Chelus* (Duméril 1806) in its large size and quadrangular carapace shape, similarly elongated anterior elements, scale radiation, and plastral shape. However, the carapace dentations and protuberances of extant *Chelus* are absent in *L. neuquina*. According to Broin and de la Fuente (1993), Lapparent de Broin and de la Fuente (2001), and de la Fuente (2007), the *Chelus* group is also recorded in Campanian–Maastrichtian horizons (Los Alamitos and Allen Formation) outcropping in other Patagonian localities (Los Alamitos, El Abra, Trapalcó). These specimens are represented by isolated cervical

[4]Courtesy of Indiana University Press.

vertebrae and shell fragments. As in the case of *Bonapartemys bajobarrealis*, Cadena and Bloch (2009) suggested that *Lomalatachelys neuquina* is neither related to *Chelus* nor Chelidae but closely related to Panpelomedusoides. However, Cadena Rueda's new phylogenetic hypothesis is simply stated, not developed, in their extended abstract.

4.1.13 *Hydromedusa* Wagler 1830

Type species: *Emys maximiliani* Mikan 1820.
Species included: *Hydromedusa casamayorensis* de la Fuente and Bona 2002, *H. maximiliani* (Mikan 1820), and *H. tectifera* Cope 1870.

Hydromedusa maximiliani (Mikan 1820)

Holotype: Naturhistorisches Museum Wien, Wien, Austria. NMW 23391.
Definition: See Ernst and Barbour (1989).
Distribution: This species is endemic to southeastern Brazilian coastal rivers from Espiritu Santo south to São Paulo (Souza and Martins 2009) (Fig. 4.8a, b).
Comments: As pointed out by Wood and Moody (1976), specimens of *Hydromedusa maximiliani* are not common in herpetological collections in comparison with specimens of *H. tectifera*. This could be a result of their restricted distribution (see Souza and Martins 2009) and/or the lesser abundance in nature rather than the actual absence of a systematic collection. A detailed osteological shell description is presented in Wood and Moody (1976) and revised data regarding the taxonomy, morphology, ecology, and population status of this species is presented by Souza and Martins (2009). According to Souza et al. (2003), the phylogeographic differentiation of *H. maximiliani* is influenced by mountain chains, and the estimated divergence time is correlated with an orogenetic event that uplifts these mountains during the Miocene and Pliocene.

Hydromedusa tectifera Cope 1870

Holotype: Not located.
Definition: See Ernst and Barbour (1989).
Distribution: Southeastern Brazil, from Goias and Minas Gerais to São Paulo; upper Parana and tributaries in Paraguay and Misiones in Northeastern Argentina; lower Paraná River, Bonaerean creek, and river of Río de la Plata basin (Argentina); and Uruguay river basin, Uruguay (Cabrera 1995, 1998) (Fig. 4.8a, c).
Comments: This species was originally described as *Hydromedusa tectifera* by Cope (1869), later referred to *Hydromedusa platanensis* by Gray (1873), and finally *Hydromedusa wagleri* by Günter (1884). Gaffney (1977) recognized derived characters in *Hydromedusa* and provides figures the skull of *H. tectifera* in several views. Most recently, *H. tectifera* was described by Cabrera (1998) who also provided figures of the skull, carapace, and plastron. The latter author summarized the natural history of this taxon.

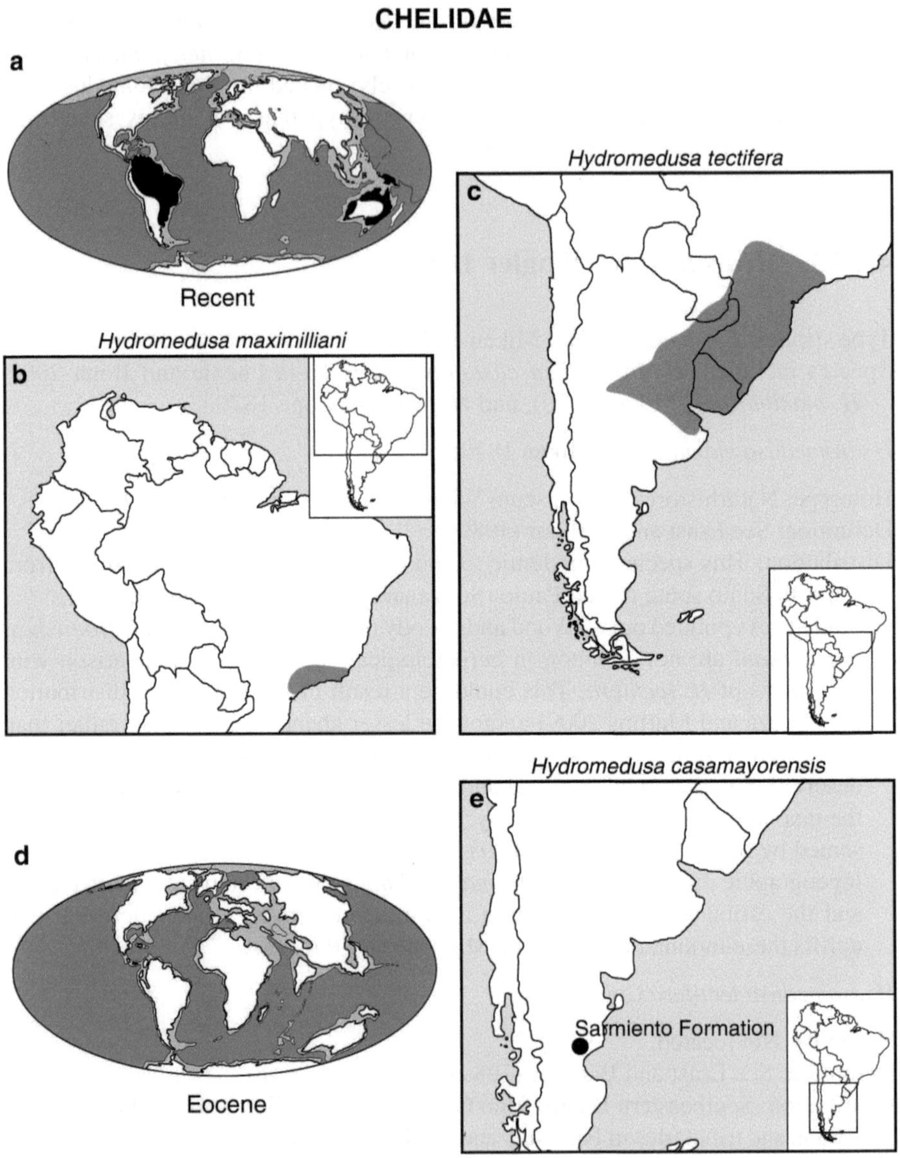

Fig. 4.8 Geographic and stratigraphic distribution of *Hydromedusa* (Chelidae). (**a**) Global distribution of Chelidae. (**b**) Geographic distribution of *Hydromedusa maximilliani* in South America. (**c**) Geographic distribution of *Hydromedusa tectifera* in South America. (**d**) Paleogeographic reconstruction of continents during the Eocene. (**e**) Geographic and stratigraphic location of *Hydromedusa casamayorensis*

Hydromedusa casamayorensis de la Fuente and Bona 2002

Holotype: Museo de Ciencias Naturales de Mar del Plata, Mar del Plata, Argentina. MCNMP 736 h, a nuchal bone.

Diagnosis: See de la Fuente and Bona (2002).

Locality and age: Cañadón Hondo area, eastern central Patagonia, Chubut Province, Argentina. Lower levels of the Sarmiento Formation, Casamayoran SALMA, Middle Eocene (Maniel et al. 2012) (Fig. 4.8d, e).

Comments: This species was named by de la Fuente and Bona (2002) on the basis of its holotype (a nuchal bone) and additional associated fragmentary material of the shell (additional material referred to this species in de la Fuente and Bona 2002). Several characters in this extinct species prompted de la Fuente and Bona (2002) to propose a specific differentiation from the extant species. These characters include the first vertebral scute, which is wide and extended onto the posteromedial extremities of the first peripheral bone, the proportion and position of the twelfth marginal scales, and the femoral/anal ratio. However, some of these characters (i.e., differences in shape and proportions of vertebral scutes) may be variable among populations of the extant *Hydromedusa tectifera* (Clavijo Baquet et al. 2010). The recent discovery of several nearly complete shells of *Hydromedusa casamayorensis* (Maniel et al. 2012) allowed these researchers to add diagnostic characters and revalidate the only extinct species of the genus *Hydromedusa*. According to the phylogenetic analyses performed by Maniel et al. (2012), *H. casamayorensis* is the sister taxa of the extant *Hydromedusa* species.

4.1.14 *Yaminuechelys* de la Fuente, Lapparent de Broin, and Manera de Bianco 2001

Type species: *Yaminuechelys gasparinii* de la Fuente, Lapparent de Broin, and Manera de Bianco 2001.

Diagnosis: See de la Fuente et al. (2001), Bona and de la Fuente (2005), and de la Fuente (2007).

Species included: *Yaminuechelys gasparinii* de la Fuente, Lapparent de Broin, and Manera de Bianco 2001 and *Y. maior* (Staesche 1929).

Yaminuechelys maior (Staesche 1929)

Neotype: Museo Paleontológico Egidio Feruglio, Trelew, Argentina. MPEF-PV 1273, posterior portion of the skull (occipital, otic, and parietal area), mandible, hyoid apparatus, atlas–axis, carapace, and plastron.

Diagnosis: See Bona and de la Fuente (2005).

Locality and age: Cerro Hansen, Chubut Province, Argentina. The horizon of the neotype and associated material was assigned to the Salamanca Formation (Lesta and Ferello 1972), Banco Negro Inferior (Feruglio 1949), and Lower Palaeocene (see Méndez 1966; Bertels 1979; Marshall et al. 1981; Somoza et al. 1995). For details of the stratigraphic section at Cerro Hansen site, see Bona et al. (1998) (Fig. 4.9a, b).

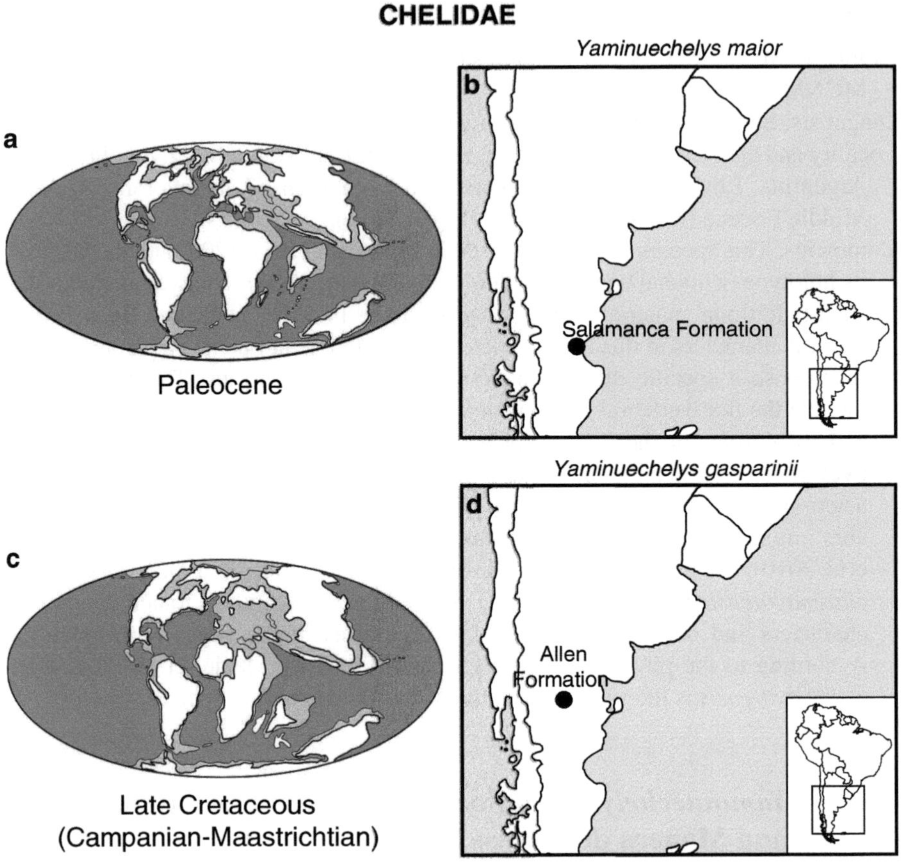

Fig. 4.9 Geographic and stratigraphic distribution of *Yaminuechelys* (Chelidae). (**a**) paleogeographic reconstruction of continents during the Palaeocene. (**b**) Geographic distribution of *Yaminuechelys maior* in South America. (**c**) Paleogeographic reconstruction of continents during the Late Cretaceous. (**d**) Geographic and stratigraphic location of *Yaminuechelys gasparinii*

Comments: The chelonian fauna of the Salamanca Formation (Lower Palaeocene) has been known since the beginning of the twentieth century, when the German paleontologist K. Staesche named several new taxa of pleurodiran and "cryptodiran" turtles in 1929.[5] These taxa were described on the basis of fragmentary shell bones recovered at Punta Peligro, Chubut Province, Argentina. Staesche (1929) indicated the presence of new species of the pleurodiran bothremydid genus *Najadochelys* (sic) Hay (1908): *N. patagonica* n. sp., *Najadochelys* cf. *patagonica* n. sp., and *?Najadochelys maior* n. sp. Likewise, Staesche (1929) recognized the presence of two "cryptodiran" species in Punta Peligro: *Osteopygis*

[5]Courtesy of The Society of Vertebrate Paleontology.

sculptus n. sp. and cf. *Gyremys* Hay 1908. Subsequent to Staesche (1929) work, Price (1953) suggested that the material described by Staesche as *Najadochelys* might have close affinities to the podocnemidid genus *Roxochelys*. Later, Wood (1971) stated that *N. patagonica* and *?N. maior* should be regarded as *nomina dubia*. Furthermore, Wood (1971, 1975 in Gasparini and Báez 1975) suggested the probable presence of some kind of pelomedusid among Staesche's fossil turtles in Punta Peligro, and de Broin (1987), Broin and de la Fuente (1993), de la Fuente (1988), and Bona et al. (1998) stated that all of Staesche's Punta Peligro species might be assigned to pleurodiran chelid turtles. Recently, new and more complete material belonging to the largest of Staesche's species was recovered at Cerro Hansen (Salamanca Formation), Chubut province (Bona and de la Fuente 2005). These materials include nearly complete skeletons with skulls and articulated cervical vertebrae that bear pleurodiran and chelid derived characters, such as the pelvis attached by suture to the shell, extended lateral emargination with absence of quadratojugal in the skull, and biconvex fifth and eighth cervical vertebrae (see Gaffney 1977; Gaffney and Meylan 1988).

Staesche (1929) assigned the largest specimens of the species that he described from the Salamanca Formation in Punta Peligro to two different taxa: *?Najadochelys maior* n. sp. (Pleurodira Bothremydidae) and *Osteopygis sculptus* n. sp. (Cryptodira, Thalassemydidae). *?Najadochelys maior* was described on the basis of fragments of two right and one left xiphiplastra (Staesche 1929: pl. XVI, Figs. 8–12), two right hyoplastral fragments, and two posterior peripheral bones (Staesche 1929: pl. XVI, Figs. 13–16). *Osteopygis sculptus* was described based on three peripherals (Staesche 1929: pl. xvii, Figs. 11–14; pl. xviii, Figs. 1–4) and two neurals (Staesche 1929: pl. xviii, Figs. 5–7). According to Bona and de la Fuente (2005), the shell bones of both species illustrated by Staesche (1929)—particularly posterior peripheral 11 and the pygal bone of ?*Najadochelys maior* (Staesche 1929: pl. xvi, Figs. 15, 16), a peripheral of the bridge, the second peripheral, and the neural bones of *Osteopygis sculptus* (Staesche 1929: pl. xvii, Figs. 11–14; pl. xviii, Figs. 3–7)—are recognizable in specimens MPEF-PV 627 (Bona and de la Fuente 2005: Fig. 6A), MPEF-PV 599 (Bona and de la Fuente 2005: Fig. 7A–B), and MPEF-PV 125 (Bona and de la Fuente 2005: Fig. 8A–D), all of which belong to a single species. According to Bona and de la Fuente (2005), these bones are similar in shape, ornamentation, and morphology. Consequently, these authors concluded that *Osteopygis sculptus* is a junior synonym of *?Najadochelys maior* and that these two large species can be assigned to a single species within the genus *Yaminuechelys*, *Y. maior* new comb. Unfortunately, Staesche's material and the museum catalogues of South American fossil collections originally housed in the Naturkunde Museum of Berlin were lost during World War II (D. Unwin, pers. comm. 2001). This circumstance forced Bona and de la Fuente (2005) to designate a neotype for *Y. maior* (Staesche 1929), following the requirements of Article 75.3.4 of the International Code of Zoological Nomenclature. *Yaminuechelys* was recently named by de la Fuente et al. (2001) on the basis of a complete specimen from the Upper Cretaceous of northern Patagonia (see above). The type species of this genus, *Y. gasparinii*, preserves the following cranial and postcranial characters: reduced, posteriorly

Fig. 4.10 Life reconstruction of *Yaminuechelys gasparinii* from the Late Cretaceous of the Allen Formation, Argentina

widening parietals; supraoccipital–squamosal contact; supraoccipital bone extending onto the posterior dorsal portion of the skull and contributing to the posterolateral margin of the cheek emarginations; quadrate participating in the temporal bar; short, wide nuchal bone and cervical scute; relatively shortened anterior peripherals; and short, wide cuneiform mesoplastra. These characters also are present in the new material of *Y. maior* species and allow Bona and de la Fuente (2005) to refer both species to a single genus. The presence of chelid species of *Yaminuechelys* in Campanian–Maastrichtian and Palaeocene rocks of Patagonia is also direct evidence of continental forms surviving the K/T mass extinction.

Yaminuechelys gasparinii de la Fuente, Lapparent de Broin, and Manera de Bianco 2001 (Fig. 4.10)

Holotype: Museo "Carlos Darwin" Punta Alta, Buenos Aires Province, Argentina. MPA 86-86-IC, carapace, plastron, skull, hyoid apparatus, atlas, axis, sixth, seventh and eight cervical vertebrae, left and right scapular girdle, left and right humeri, right radius and ulna, pelvic girdle, left and right femora, right tibia and fibula, and two sacral and seven caudal vertebrae.

Diagnosis: See de la Fuente et al. (2001), Bona and de la Fuente (2005), de la Fuente (2007).

Locality horizon and age: This turtle was found in a mudstone level belonging to a succession of mudstones and sandstones designated lacustrine Senonian by Wichmann (1927), exposed at Yaminué Creek, Cerro Blanco, Río Negro Province, Argentina (Manera de Bianco 1996). These sediments are similar to those of the Lago Pellegrini area, which are assigned (Andreis et al. 1974) to the middle Member of the Allen Formation, Upper Cretaceous (Upper Campanian–Lower Maastrichtian; Hugo and Leanza 2001a; Leanza et al. 2004) (Fig. 4.9c, d).

Comments: *Yaminuechelys gasparinii* holotype was discovered by Dr. Teresa Manera de Bianco. *Yaminuechelys* was recovered as the sister group of *Hydromedusa*, supported by the large *aperturae narium internae* with reduced ossification of palatines (Bona 2004; Bona and de la Fuente 2005; de la Fuente 2007). Long-necked chelid remains assigned to the *Hydromedusa* subgroup, probably belonging to *Yaminuechelys,* have been found in several localities where horizons of Santonian–Maastrichtian age crop out (de Broin 1987; Broin and de la Fuente 1993; Gasparini and de la Fuente 2000; de la Fuente et al. 2001 and references therein, de la Fuente 2007; de la Fuente et al. 2010; Maniel and Apesteguía 2011).

4.2 Discussion

Chelidae is a monophyletic group with extant species distributed across Australasia and South America (Gaffney et al. 2006; Pritchard and Trebbau 1984; de Broin 1988; Iverson 1992). Since the end of the nineteenth century, two groups of extant chelid turtles (short and long necked) have been recognized by European zoologists (e.g., Boulenger 1989) based on the length of the eight cervical vertebrae in comparison with the length of the ten thoracic vertebrae attached to the carapace. According to de la Fuente et al. (2010), aside from the difference in the length of cervical vertebra, there are other major differences between the cervical vertebrae of short- and long-necked chelids, e.g., variation in the angle between the axis of the process that supports the postzygapophyses and the cervical vertebral centrum and the existence of contact (or not) between the postzygapophyses in each cervical vertebra (Goody 1967). Although phylogenies of the short- and long-necked chelids have been hinted at in different sources, there was little agreement among them resulting in conflicting hypotheses. These phylogenies have been based on morphology (Burdbidge et al. 1974; Gaffney 1977; McDowell 1983; Gaffney and Meylan 1988; Bona and de la Fuente 2005), karyotypes (Bull and Legler 1980), serology (Frair 1962, 1980; Burdbidge et al. 1974), electrophoresis (Georges and Adams 1992), molecular data (Seddon et al. 1997; Georges et al. 1998), and bone microstructure (Scheyer 2009). The main phylogenetic hypothesis could be summarized by two competing hypotheses: the classical morphological approach and the molecular and serological analyses. The morphological phylogenies based on cranial and postcranial characters concluded that the long-necked chelids, *Chelodina* (Australia), *Hydromedusa* (South America), and *Chelus* (South America), form a monophyletic group spanning in both continents (Gaffney 1977; Gaffney and Meylan 1988; Bona and de la Fuente 2005). To the contrary, based on morphological and serological analysis (Burdbidge et al. 1974) and on molecular data (Seddon et al. 1997; Shaffer et al. 1997; Georges et al. 1998; Fujita et al. 2004; Near et al. 2005; Krenz et al. 2005), these authors concluded that the Australian short- and long-necked species are more closely related to each other than to the South American species (Fig. 4.11). The molecular analyses suggest an independent radiation of the chelids following the separation of the Australian and South American continents and contradict the monophyly of long-necked chelids proposed by Gaffney (1977), Gaffney and Meylan (1988), and Bona and de la Fuente (2005). According to Bona and de la Fuente (2005), *Chelodina*, *Hydromedusa*, and

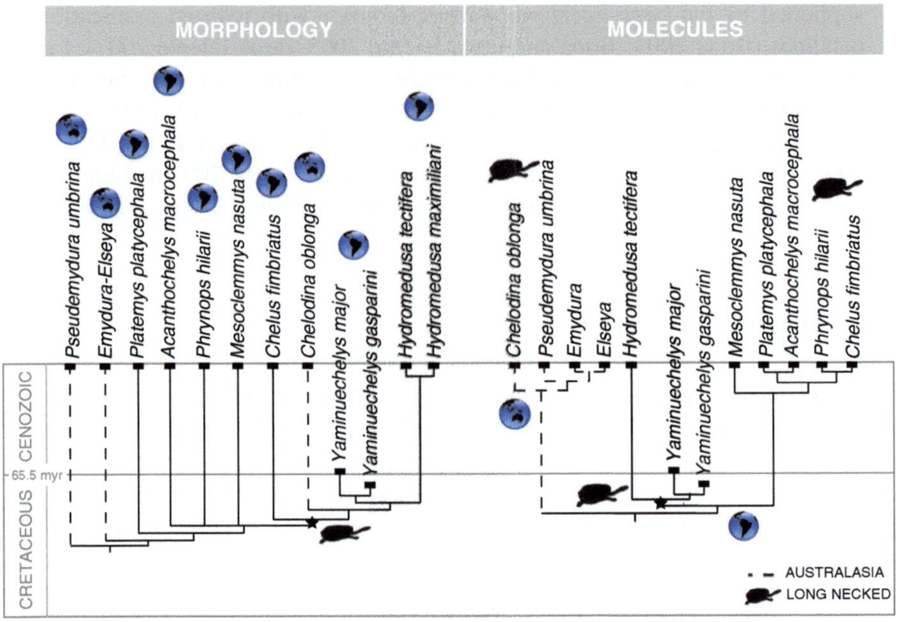

Fig. 4.11 Calibrated trees of competing hypotheses of chelid evolution. *Left cladogram*, hypothesis resulted from morphological data (e.g., Gaffney 1977; Gaffney and Meylan 1988; Bona and de la Fuente 2005). *Right cladogram*, hypothesis resulted from molecular data (e.g., Seddon et al. 1997; Shaffer et al. 1997; Georges et al. 1998; Fujita et al. 2004; Near et al. 2005; Krenz et al. 2005)

Yaminuechelys represent extant and extinct genera of a long-necked lineage that were differentiated prior to the breakup of the southern Gondwana landmass. Furthermore, the occurrence of isolated cervical vertebrae with similar morphology to that of some extant Australasian short-necked chelids (e.g., *Emydura*, *Elseya*) in sediments of Campanian–Maastrichtian age also supports the proposed statement (Broin and de la Fuente 1993, plate 1, 2–3).

The oldest chelids are known from the Aptian–Albian in Patagonia (Lapparent de Broin and de la Fuente 2001; de la Fuente 2007; de la Fuente et al. 2011) and from the Albian in Australia (Smith 2009, 2010), as well as the Oligocene in Tasmania (Warren 1969). The present disjoined distribution, the early fossil record of this clade, and several peculiarities of extant chelids noted by Pritchard (1984) (e.g., that tropical chelids have an amazing tolerance to cold temperatures and, contrary to other reptiles, chelids exhibit great diversity toward the southern extreme of their range) suggest a much older biogeographical and phylogenetic history on southern Gondwana landmasses. The paleogeography of southern Gondwana has a complex history that was summarized by Woodburne and Case (1996). Geophysical and geological evidence suggests that at least as early as the Late Cretaceous, the Antarctic Peninsula was yet contiguous with southern South America (Grunow 1992; Lawver et al. 1992; Shen 1995). These data are consistent with the occurrence of other tetrapods, including monotremes, in the Early Palaeocene faunas of Patagonia (Pascual et al. 1992) and ratites in Late Eocene faunas of Seymour Island (Antarctic Peninsula) (Tambussi et al. 1994). The occurrence of these tetrapods in Patagonia, the Antarctic Peninsula, and Australia might be the result of a much older, widespread southern Gondwana distribution of these taxa prior to the Late Cretaceous.

References

Andreis RR, Iñíguez Rodríguez AM, Lluch JJ, Sabio DA (1974) Estudio sedinentológico de las formaciones del Cretácico superior del Lago Pellegrini (provincia de Río Negro, República Argentina). Rev Asoc Geol Arg 29:85–102

Barcat C, Cortinas JS, Nevitic VA, Zucchi HE (1989) Cuenca Golfo de San Jorge. In: Chebbli G, Spalletti LA (eds) Cuencas Sedimentarias Argentinas, vol 6, Serie Correlación Geológica. Instituto Superior de Correlación Geológica, Universidad Nacional de Tucumán, San Miguel de Tucumán

Bertels A (1979) Paleobiogeografía de los foraminíferos del Cretácico Superior y Cenozoico de América del Sur. Ameghiniana 16:273–356

Bona P (2004) Sistemática y biogeografía de las tortugas y los cocodrilos paleocenos de la Formación Salamanca, Provincia de Chubut, Argentina. Doctoral dissertation thesis, Universidad Nacional de La Plata, La Plata

Bona P (2006) Paleocene (Danian) chelid turtles from Patagonia, Argentina: taxonomic and biogeographic implications. N Jahr Geol Palaontol 241:303–323

Bona P, de la Fuente MS (2005) Phylogenetic and paleobiogeographic implications of *Yaminuechelys maior* (Staesche, 1929) new comb. A large long-necked chelid turtle from the Early Paleocene of Patagonia. J Vertebr Paleontol 25:569–582

Bona P, Cladera G, de la Fuente MS (1998) Las tortugas pleurodiras de la Formación Salamanca (Paleoceno inferior) en el área de Cerro Hansen, Provincia de Chubut, Argentina. 10° Congr Latin Geol 6° Congr Nac Geol Económica 1:269–274

Bonaparte CLJL (1836) Chcloniorum Tahula Analytica, Rome

Bonaparte JF (1990) New Late Cretaceous mammals from the Los Alamitos Formation, northern Patagonia. Nat Geogr Res 6:63–93

Bonaparte JF (1991) Los vertebrados fósiles de la Formación Río Colorado, de la ciudad de Neuquén y cercanias, Cretácico Superior, Argentina. Rev Mus Arg Cs Nat "Bernardino Rivadavia", Paleontología 4:15–123

Bonaparte JF, Franchi MR, Powell JE, Sepúlveda EG (1984) La Formación Los Alamitos (Campaniano-Maastrichtiano) del sudeste de Río Negro, con descripción de *Kritosaurus australis* n. sp. (Hadrosauridae). Significado paleogeográfico de los vertebrados. Rev Asoc Geol Arg 39:284–299

Boulenger GA (1989) Catalogue of the Chelonians, Rhynchocephalians, and Crocodiles in the British Museum (Natural History). Trustees of the British Museum (Natural History), London

Bour R (1973) Contribution à la connaissance de *Phrynops nasutus* (Schweiggcr: 1812) et *Phrynops tuberculatus* (Luederwaldt 1926). Description d'une nouvelle sous-espéce originaire du Paraguay, *Phrynops tuberculatus vanderhaegei*. (Testudinata-Pleurodira-Chelidae). Bull Soc Zool Fr 98:175–190

Bour R, Pauler I (1987) Identité de *Phrynops vanderhaegei* Bour, 1973, et des espéces affines (Reptilia-Chelonii-Chelidae). Mesogée 47:3–23

Bour R, Zaher H (2005) A new species of *Mesoclemmys*, from the open formations of Northeastern Brazil (Chelonii: Chelidae). Pap Avul Zool 45:295–311

Bridge J, Jalfin G, Georgieff S (2000) Geometry, lithofacies and special distribution of Cretaceous fluvial sandstone bodies, San Jorge Basin, Argentina, outcrop analog for the hydrocarbon-bearing Chubut Group. J Sed Res 70:341–359
Bull J, Legler J (1980) Karyotypes of side-necked turtles (Testudines: Pleurodira). Can J Zool 58:828–841
Burdbidge AA, Kirsch JA, Main AR (1974) Relationships within the Chelidae (Testudines: Pleurodira) of Australia and New Guinea. Copeia 1974:392–409
Cabrera MT (1995) Comparative composition of turtles species in four natural regions of the Chacoan Domain, South America. An Mus Hist Nat Valparaíso 23:41–52
Cabrera MT (1998) Las Tortugas Continentales de Sudamérica Austral. BR Copias, Córdoba
Cadena E, Bloch J (2009) Two new specimens of the platychelyid turtle Notoemys zapatacaensis from the Early Cretaceous (Valanginian) of Colombia. Spec Publ Royal Tyrrell Museum Turtle Symposium Abstracts and Programs 1:29–31
Cann J, Legler J (1994) The Mary River tortoise a new genus and species of short-necked chelid from Queensland, Australia (Testudines: Pleurodira). Chelon Conserv Biol 1:81–96
Cione A, Tonni EP (1995) Chronoestratigraphy and "Land mammal ages" in the Cenozoic of southern South America: principles, practices, and the "Uquian" problem. J Paleontol 69:135–159
Cione AL, Azpelicueta M, Bond M, Carlini A, Casciotta R, Cozzuol M, de la Fuente MS, Gasparini Z, Goin J, Noriega J, Scillato-Yané J, Soibelzon L, Tonni EP, Verzi D, Vucetich MG (2000) Miocene vertebrates from Paraná, eastern Argentina. Correlación Geol 14:191–237
Clavijo Baquet S, Loureiro M, Achaval F (2010) Morphological Variation in the South American Snake-Necked Turtle *Hydromedusa tectifera* (Testudines: Chelidae). Chelon Conserv Biol 9:231–237
Cope ED (1869) Seventh contribution to the herpetology of tropical America. Proc Am Philos Soc 11:147–169
de Broin F (1987) The Late Cretaceous Fauna of Los Alamitos, Patagonia, Argentina. Part IV, Chelonia. Rev Mus Arg Cien Nat 'Bernardino Rivadavia' Paleontol 3:131–139
de Broin F (1988) Les Tortues et le Gondwana. Examen des rapport entre le fractionemment du Gondwana et la dispersión géographique dess tortues pleurodires à partir de Crétacé. Stud Geol Salmant Stud Palaeochel 2:103–142
de Broin F, de la Fuente MS (1993) Les tortues fossiles d'Argentine: synthèse. Ann Paléontol 79:169–232
de Gasparini ZB, Báez AM (1975) Aportes al conocimiento de la herpetofauna terciaria de la Argentina. Actas 1° Congr Arg Paleont Bioestr Tucumán 2:377–415
de Gasparini ZB, de la Fuente MS (2000) Tortugas y plesiosaurios de la Formación La Colonia (Cretácico Superior) de Patagonia. Rev Esp Paleontol 15:23–35
de Gasparini ZB, de la Fuente MS, Donadio O (1986) Los Reptiles Cenozoicos argentinos: Implicancias paleoambientales y evolución biogeográfica. Actas 4° Congr Arg Paleontol Bioestr 2:119–130
de la Fuente MS (1986) Un quélido Montehermosense (Plioceno temprano-medio) en las proximidades de Cósquin, Provincia de Córdoba, Argentina. Ameghiniana 23:65–74
de la Fuente MS (1988) Las tortugas Chelidae (Pleurodira) y Testudinidae (Cryptodira) del Cenozoico argentino. Doctoral dissertation, Universidad Nacional de La Plata, La Plata
de la Fuente MS (1992) Las tortugas Chelidae del Terciario superior y Cuaternario del territorio argentino. Ameghiniana 29:211–299
de la Fuente MS (2003) Two new Pleurodiran turtles from the Portezuelo Formation (Upper Cretaceous) of northern Patagonia, Argentina. J Paleontol 77:559–575
de la Fuente MS (2007) Testudines. In: de Gasparini BZ, Coria R, Salgado L (eds) Patagonian Mesozoic reptiles. Indiana University Press, Bloomington
de la Fuente MS, Bona P (2002) Una nueva especie del género *Hydromedusa* Wagler (Pleurodira Chelidae) del Paléogeno de Patagonia. Ameghiniana 39:77–83
de la Fuente MS, Cabrera M (1988) Comentarios sobre las tortugas fósiles de Sierra Chica (provincia de Córdoba, Argentina). Bol Asoc Herpetol Arg 4:65–74
de la Fuente MS, Ledesma A (1985) Presencia de *Platemys* sp. en sedimentitas del Neógeno de la provincia de Córdoba, Argentina. Bol Asoc Herpetol Arg 2:9

de la Fuente MS, Lapparent de Broin F, Manera de Bianco T (2001) The oldest and first nearly complete skeleton of a chelid, of the *Hydromedusa* sub-group (Chelidae, Pleurodira), from the Upper Cretaceous of Patagonia. Bull Soc Géol Fr 172:237–244

de la Fuente MS, Calvo J, González Riga BJ (2007) A new Cretaceous chelid turtle from Northern Neuquén Basin, Neuquén Province, Argentina. Ameghiniana 44:485–492

de la Fuente MS, Barbieri R, Chafrat P (2010) Una tortuga Chelidae (Testudines: Pleurodira) de cuello largo en el Grupo Neuquén, Río Negro, Argentina. Significado cronológico y paleobiogeográfico. Andean Geol 37:398–412

de la Fuente MS, Umazano AM, Sterli J, Carballido JI (2011) New chelid of the lower section of the Cerro Barcino formation (Aptian-Albian?), Patagonia. Cretaceous Res 32:527–537

Derr JN, Bickham JW, Greebaum IF, Rhodin AGJ, Mittermeier RA (1987) Biochemical systematics and evolution in the South American turtle *Platemys* (Pleurodira: Chelidae). Copeia 1987:370–375

Duméril AMC (1806) Zoologie analytique au methode naturelle de classification des animaux, rendue plus facile à l'aide de tableaux synoptiques. Allais, Paris

Duméril AMC, Bibron G (1835) Erpetologie generale ou histoire naturelle complete des reptiles, vol 2. Librairie Encyclopedique de Roret, Paris

Ernst CH (1983a) Platemys pallidipectoris. Cat Am Amphib Rept 325:1–2

Ernst CH (1983b) Platemys spixii. Cat Am Amphib Rept 326:1–2

Ernst CH (1983c) Platemys radiolata. Cat Am Amphib Rept 339:1–2

Ernst CH (1984) Geographic variation in the Neotropical turtle, *Platemys platycephala*. J Herpetol 17:345–355

Ernst CH (1987) Platemys platycephala. Cat Am Amphib Rept 405:1–4

Ernst CH, Barbour RW (1989) Turtles of the world. Smithsonian Institution Press, Washington, DC

Feruglio E (1949) Descripción geológica de la Patagonia. Yac Petrol Fis 2:1–349

Frair W (1962) Turtle family relationships as determined by serological tests. In: Leone CA (ed) Taxonomic biochemistry and serology. Ronald Press, New York

Frair W (1980) Serological survey of Pleurodiran turtle. Comp Biochemist Physiol 65:505–511

Frair W (1982) Serological studies of the red turtle *Phrynops rufipes*. Herp Bull New York Herpetol Soc 17:4–9

Freiberg MA (1945) Una nueva especie de tortuga del genero *Platemys* Wagler. Physis 20:19–23

Fujita MK, Engstrom TN, Starkey DE, Shaffer HB (2004) Turtle phylogeny: insights from a novel nuclear intron. Mol Philogenet Evol 31:1031–1040

Gaffney ES (1977) The side-necked turtle family Chelidae: a theory of relationships using sharing characters. Am Mus Novit 2620:1–28

Gaffney ES, Meylan PA (1988) A phylogeny of turtles. In: Benton MJ (ed) The phylogeny and classification of the tetrapods 1: amphibians, reptiles, birds, vol 35A, Systematics association, special volume. Clarendon, Oxford

Gaffney ES, Tong H, Meylan PA (2006) Evolution of the side-necked turtles: the families Bothremydidade, Euraxemydidae, and Araripemydidae. Bull Am Mus Nat Hist 300:1–698

Georges A, Adams M (1992) A phylogeny for Australian chelid turtles based on allozyme electrophoresis. Aust J Zool 40:453–476

Georges A, Birrell J, Saint KM, McCord WP, Donnellan SC (1998) A phylogeny for side-necked turtles (Chelonia: Pleurodira) based on mitochondrial and nuclear gene sequence variation. Biol J Linn Soc 67:213–246

Goody J (1967) Freshwater tortoises of Australia and New Guinea (in the family Chelidae). Lansdowne Press, Melbourne

Gray JE (1844) Catalogue of the tortoises, crocodiles, and amphisbaenians in the collection of the British Museum. British Museum (Natural History), London

Gray JE (1855) Catalogue of shield reptiles in the collection of the British Museum. Part I. Testudinata I (tortoises). Taylor & Francis, London

Gray JE (1867) Description of a new Australian tortoise *(Elseya latisternum)*. Ann Mag Nat Hist Lond 20:43–45

Gray JE (1872) Appendix to the catalogue of shield reptiles in the collection of the British Museum. Part I. Testudinata (tortoises). British Museum, London

Gray JE (1873) Observations on Chelonian, with description of a new genera and species. Ann Mag Nat Hist 11:289–308

Groombridge B (1982) The IUCN Amphibia-Reptilia Red Data Book. Testudines, Crocodylia, Rhynchocephalia Part 1. IUCN, Gland

Grunow A (1992) Creation and destruction of Wedell Sea floor in the Jurassic. Geology 21:647–650

Günter A (1884) Contribution to our knowledge of *Hydromedusa*, a genus of South American freshwater turtles. An Mag Nat Hist 14:421–425

Hay OP (1908) The fossil turtles of North America. Carnegie Inst Wash Publ 75:1–568

Hoogmoed MS, Grüber U (1983) Spix and Wagler type specimens of reptiles and amphibians in the Natural History Museum in Munich (Germany) and Leiden (The Netherlands). Spixiana Suppl 9:319–415

Hugo CA, Leanza HA (2001a) Hoja Geológica 3966-III, Villa Regina, Provincia de Río Negro. Inst Geol Rec Nat Bol Serv Geol Min Arg 309:1–53

Hugo CA, Leanza HA (2001b) Hoja Geológica 3969-IV, General Roca, provincias de Neuquén y Río Negro. Inst Geol Rec Nat Bol Serv Geol Min Arg 308:1–71

Iverson JB (1986) A checklist with distribution maps of the turtles of the world. Iverson, Richmond

Iverson JB (1992) A revised checklist with distribution maps of the turtles of the world. Privately Printed, Richmond

King FW, Burke RL (1989) Crocodilian, tuatara and turtle species on the world. Association of Systematics Collections, Washington, DC

Kischlat EE (1993) Contribuçao ao conhocimento de quélidas fósseis e recents do Brasil Quélidas (Chelonii, Pleurodira) da Bacia de Tabuaté Cenozoico do Estado do São Paulo, Brasil. Magister dissertation thesis, Universidade Federal do Rio de Janeiro, Rio de Janeiro

Krenz JG, Naylor GJP, Shaffer HB, Janzen FJ (2005) Molecular phylogenetics and evolution of turtles. Mol Phylogenet Evol 37:178–191

Lamar WW, Medem F (1984) Notes on the chelid turtle *Phrynops rufipes* in Colombia (Reptilia: Testudines: Chelidae). Salamandra 18:305–321

Lapparent de Broin F, de la Fuente MS (2001) Oldest world Chelidae (Chelonii, Pleurodira), from the Cretaceous of Patagonia. C Rend Acad Sci Paris 333:463–470

Lawver LA, Gahagan LM, Coffin FM (1992) The development of paleoseaways around Antarctica. Am Geophys Union Antarc Res Ser 65:7–30

Leanza HA (1999) The Jurassic and Cretaceous terrestrial beds from Southern Neuquén Basin, Argentina. Field Guide Inst Sup Correl Geol Miscelánea 4:1–30

Leanza HA, Hugo CA (2001) Cretaceous red beds from southern Neuquén Basin (Argentina): age, distribution and stratigraphic discontinuities. 7° International Symposium Mesozoic Terrestrial Ecosystems Asoc Paleont Arg Publ Esp 7:111–122

Leanza HA, Apesteguía S, Novas FE, de la Fuente MS (2004) Cretaceous terrestrial beds from Neuquén Basin (Argentina) and their tetrapods assemblages. Cretaceous Res 25:61–87

Legler JM, Cann J (1980) A new genus and species of chelid turtle from Queensland, Australia. Contrib Sci Nat Hist Mus Los Angel Co 324:1–18

Lesta P, Ferello R (1972) Región extraandina de Chubut y Norte de Santa Cruz. In: Leanza A (ed) Geología Regional Argentina. Academia Nacional de Ciencias, Córdoba

Lindholm WA (1929) Revidierttes Verzeichnis der Gattungen der rezenten Schildkröten nebst Notizen zur Norre1klatur einiger Arten. Zool Anz 81:275–295

Luederwaldt H (1926) Os Chelonios brasileiros com a lista das especies do Museu Paulista. Rev Mus Paulista 14:404–468

Manera de Bianco T (1996) Nueva localidad con nidos de huevos de dinosaurios (Titanosauridae) del Cretácico de Cerro Blanco, Yaminué, Río Negro, Argentina. Asoc Paleont Arg Publ Esp 4:59–67

Maniel IJ, Apesteguía S (2011) Tortugas Pleurodiras de la Formación Bajo de la Carpa (Santoniano) de ¨La Bonita¨, Provincia de Río Negro (Argentina). Implicaciones Biogeográficas. 4° Congr Lat Paleont Vertebr, San Juan, Provincia de San Juan

Maniel IJ, de la Fuente MS, Sterli J (2012) New Material of *Hydromedusa casamayorensis* from the Middle Eocene of Patagonia and its Phylogenetic Implications. Symposium on turtle evolution, Tübingen

Marshall LG, Butler RF, Drake RE, Curtis GH (1981) Calibration of the beginning of the age of mammals in Patagonia. Science 212:43–45

Martínez R, Giménez O, Rodríguez J, Bochatey G (1986) *Xenotarsosaurus bonapartei* nov. gen. et sp. (Carnosauria, Abelisauridae), un nuevo Theropodo de la Formación Bajo Barreal, Chubut, Argentina. Actas 4° Congr Arg Paleont Bioestr 2:23–32

McBee K, Bickham JW, Rhodin AGJ, Mittermeier RA (1985) Karyotypic variation in the genus Platemys (Testudines: Pleurodira). Copeia 1985:445–449

McCord WP, Joseph-Ouni M, Lamar W (2001) Taxonomic reevaluation of *Phrynops* (Testudines: Chelidae) with the description of two new genera and a new species of *Batrachemys*. Rev Biol Tropical 49:715–764

McDowell SB (1983) The genus *Emydura* (Testudines: Chelidae) in New Guinea with notes on the penial morphology of Pleurodira. In: Rhodin AGJ, Miyara K (eds) Advances in herpetology and evolutionary biology: essays in honor of Ernest E Williams. Museum of Comparative Zoology, Harvard University, Cambridge

Medem F (1960) Informes sobre Reptiles colombianos (IV). El primer hallazgo de la tortuga *Phrynops* (*Batrachemys*) *nasuta* (Schweigger) en Colombia. Novededades Colombianas 5:284–290

Medem F (1973) El primer hallazgo de la tortuga *Phrynops rufipes* (Spix) en Colombia. Rev Acad Col Cienc 4:49–68

Méndez I (1966) Foraminíferos, edad y correlación estratigráfica del Salamanquense de Punta Peligro (45° 30'S; 67° 11'W) Provincia del Chubut. Rev Asoc Geol Arg 21:127–157

Merrem B (1820) Versuch eines Systems der Amphibien, Tentamen Systematic Amphiborum. Krieger, Marburg

Mertens R (1967) Bemerkenswerte Susswasserschildkröten aus Brasilien. Senck Biol 48:71–82

Mertens R (1969) Eine neue Halswender-Schildkrote aus Peru. Senck Biol 50:132

Mikan JC (1820) Delectus Florae et Faunae Brasiliensis. Antoinii Strauss, Wien

Mittermeier RA, Medem F, Rhodin AGJ (1980) Vernacular names of South American turtles. Soc Stud Amph Rept Herpetol Circ 1:1–44

Mocelin MA, Fernandes R, Porto M, Fernandes DC (2008) Reproductive biology and notes on natural history of the side-necked turtle *Acanthochelys radiolata* (Mikan 1820) in captivity (Testudines: Chelidae). S Am J Herp 3:223–228

Near TJ, Meylan PA, Shaffer HB (2005) Assessing concordance of fossil calibration points in molecular clock studies: an example using turtles. Am Nat 165:137–146

Pascual R, Archer M, Ortiz Jaureguizar E, Prado E, Godthelp H, Hand SJ (1992) First discovery of monotremes in South America. Nature 356:704–706

Perea D, Ubilla M, Piñeiro G (1996) First Fossil Reccord (Late Miocene) of *Phrynops* (*P. geoffroanus*) from Uruguay: bioestratigraphical and Palaeoenvironmental Context. Copeia 1996:445–451

Price LI (1953) Os Quelônios da Formaçao Bauru, Cretaceo terrestre do Brasil meridional. Bol Div Geol Mineral Depart Nac Prod Min 147:1–34

Pritchard PCH (1979) Encyclopedia of turtles. Tropical Fish Hobbyist Publications, Neptune

Pritchard PCH (1984) Evolution and zoogeography of South American turtles. Studia Geol Salmant Vol Esp 1 Stud Palaeochel 1:225–233

Pritchard PCH (1988) A survey of neural bone variation among recent chelonian species, with functional interpretations. Acta Zool Cracov 31:625–686

Pritchard PCH (2008) *Chelus fimbriata* (Schneider 1783) matamata turtle. In: Rhodin AGJ, Pritchard PCH, van Djik PP, Saumure RA, Buhlman KA, Iverson JB, Mittermeier RA (eds) Conservation biology of freshwater turtles and tortoises: a compilation project of the IUCN/SSC tortoises and freshwater turtle specialist group, vol 5, Chelonian research monographs. Chelonian Research Foundation, Lunenburg, pp 020.1–020.10

Pritchard PCH, Trebbau P (1984) The turtles of Venezuela. Contrib Herpetol 2:1–403

Reed KM, Hanks BG, Bickham W, Rhodin AGJ, Greenhaum IF, Mittermeier RA, Fedullo LP (1991) Cytogenetic analysis of the plcurodire turtle *Phrynops hogei* and its taxonomic implications. Amphibia-Reptilia 12:203–212

Rhodin AGJ (1985) Status and taxonomic relationships of South American chelid turtles. Am Phil Soc Grant Rep 1984:291–293

Rhodin AGJ, Mittermeier RA (1983) Description of *Phrynops williamsi* a new species of chelid tunle of the South American *P. geoffroanus* complex. In: Rhodin AGJ, Gand A, Miyata K (eds) Advances in herpetology and evolutionary biology. Museum of Comparative Zoology, Harvard University, Cambridge

Rhodin AGJ, Mittermeier RA, McMorris JR (1984) *Platemys macrocephala* a new species of chelid turtle from central Bolivia and the Pantanal region of Brazil. Herpetologica 40:38–46

Rhodin AGJ, Métrailler S, Vinke T, Vinke S, Artner H, Mittermeier RA (2009) *Acanthochelys macrocephala* Rhodin, Mittermeier, and McMorris 1984—big headed Pantanal swamp turtle. In: Rhodin AGJ, Pritchard PCH, van Djik PP, Saumure RA, Buhlman KA, Iverson JB, Mittermeier RA (eds) Conservation biology of freshwater turtles and tortoises: a compilation project of the IUCN/SSC tortoises and freshwater turtle specialist group, vol 5, Chelonian research monographs. Chelonian Research Foundation, Lunenburg, pp 040.1–040.8

Rhodin AGJ, Van Dijk PP, Iverson JB, Broadley Shaffer J (1983) Turtles of the world, 2010 update: annoted checklist of taxonomy, synonymy, distribution, and conservation status. In: Rhodin AGJ, Pritchard PCH, van Djik PP, Saumure RA, Buhlman KA, Iverson JB, Mittermeier RA (eds) Conservation biology of freshwater turtles and tortoises: a compilation project of the IUCN/SSC tortoises and freshwater turtle specialist group, vol 5, Chelonian research monographs. Chelonian Research Foundation, Lunenburg, pp 000.85–000.164

Richard E, de la Fuente MS (1992) Lista sistemática y distribución de de las tortugas argentinas (Reptilia: Chelonii). Acta Zool Lilloana 41:357–364

Sánchez Villagra MR, Pritchard PCH, Paolillo A, Linares OJ (1995) Geographic variation in the matamata turtle, *Chelus fimbriatus*, with observations on its shell morphology and morphometry. Chelon Cons Biol 1:293–300

Scheyer T (2009) Conserved bone microstructure in the shell of long-necked and short-necked chelid turtles (Testudinata, Pleurodira). Foss Rec 12:47–57

Schneider JG (1783) Allgemeine Naturgeschichte der Schildkröten, nebst einem Systematischen Verseichnisse der einzelnen Arten und zwei Kupfern. Müller, Leipzig

Schneider JG (1792) Beschreibung und Abbildung einer neun Art von Wasserschildkröten nebst Bestimmungen einiger bisher wenig bekannten fremden Arten. Schrift Ges Nat Fruende Berl 10:259–283

Schweigger AF (1812) Prodomus Monographiae Cheloniorum. Königsberg Arch Naturwiss Math 1:171–368

Seddon JM, Georges A, Baverstock PR, McCord W (1997) Phylogenetic relationship of chelid turtles (Pleurodira: Chelidae) based on mitochondrial 12S rRNA sequence variation. Mol Phylogent Evol 7:55–61

Shaffer HB, Meylan P, Mcknight ML (1997) Test of turtle phylogeny: molecular, morphological, and paleontological approaches. Syst Biol 46:235–268

Shen Y (1995) A paleoisthmus between southern South America and Antarctic Peninsula during Late Cretaceous and Early Tertiaty. In: 7th international symposium on Antarctic earth sciences, Siena. Abstracts: 345

Siebenrock F (1901) Beschreibung einer neuen Schildkrötengattung aus der Familie Chelydidae von Australien: *Pseudemydura*. Anz Akad Wiss Wien Math Nat 38:248–250

Smith ET (2009) Turtle treasures from Opal fields of Lighting ridge, New SouthWales. Turtle Simposiun. In: Braman D, Brinkman D, Marion C (eds) Abstracts and program, vol 1, Special publication of the Royal Tyrell Museum. Royal Tyrrell Museum of Paleontology, Drumheller, pp 167–168

Smith ET (2010) Early Cretaceous chelids from Lightning Ridge. New South Wales Alcheringa 34:375–384

Somoza R, Cladera G, Archangelsky S (1995) Una nueva tafoflora paleocena de Chubut, Patagonia. Su edad y ambiente de depositación. 6° Congr Arg Paleontol Bioestr Actas 1995:265–269

Souza FL, Martins FI (2009) *Hydromedusa maximiliani* (Mikan 1825)-Maximilian´s Snake Turtle, Brazilian-Necked Turtle. In: Rhodin AGJ, Pritchard PCH, van Djik PP, Saumure RA, Buhlman KA, Iverson JB, Mittermeier RA (eds) Conservation biology of freshwater turtles and tortoises: a compilation project of the IUCN/SSC tortoises and freshwater turtle specialist group, vol 5, Chelonian research monographs. Chelonian Research Foundation, Lunenburg, pp 026.1–026.6

Souza FL, Cunha AF, Oliveira MA, Pereira GAG, dos Reis SF (2003) Preliminary phylogeographic analysis of the Neotropical freshwater turtle *Hydromedusa maximiliani* (Chelidae). J Herpetol 37:427–433

Staesche K (1929) Schildkrötenreste aus der oberen Kreide Patagoniens. Paläontographica 72:103–112

Stejneger L (1909) Generic name of some Chelyid turtles. Proc Biol Soc Wash 15:125–128

Tambussi C, Noriega J, Gazdzicki A, Tatur A, Reguero MA, Vizcaino S (1994) Ratite bird from the Paleogene La Meseta formation, Seymour Island, Antarctica. Polish Polar Res 15:15–20

Thomson S, Georges A (2009) *Myuchelys* gen. nov.—a new genus for *Elseya latisternum* and related forms of Australian freshwater turtle (Testudines, Pleurodira, Chelidae). Zootaxa 2053:32–42

Vargas-Ramírez M, Michels J, Castaño-Mora OV, Cárdenas-Arevalo G, Gallego-García N, Frits U (2012) Weak genetic divergence between the two South American toad-headed turtles *Mesoclemmys dahli* and *M. zuliae* (Testudines: Pleurodira: Chelidae). Amphibia-Reptilia 33:373–385

Vinke T, Vinke S, Richard E, Cabrera M, Paszko L, Marano P, Métrailler S (2011) *Acanthochelys pallidipectoris* (Freiberg 1945) Chaco side-Necked Turtle. In: Rhodin AGJ, Pritchard PCH, van Djik PP, Saumure RA, Buhlman KA, Iverson JB, Mittermeier RA (eds) Conservation biology of freshwater turtles and tortoises: a compilation project of the IUCN/SSC tortoises and freshwater turtle specialist group, vol 5, Chelonian research monographs. Chelonian Research Foundation, Lunenburg, pp 065.1–065.7

Volkheimer W, Gallego OF, Cabaleri NG, Armella C, Narváez PL, Silva Nieto DG, Páez MA (2009) Stratigraphy, palynology and conchostracans of a Lower Cretaceous sequence at the Cañadón Calcáreo locality, extra-andean central Patagonia: age and paleoenvironmental significance. Cretaceous Res 30:270–282

von Spix JB (1824) Animalia nova species novae Testudinum et Ranarum quas in itinere per Brasiliam annis 18I 7–1820 collegit et descripsit. Hubschmann, Munchen

Wagler J (1830) Natürliches System der Amphibien mit Vorangehander Classification der Säugthiere und Vogel. J G Cottaschen Buchhandlung, München

Warren JW (1969) Chelid turtles from the Mid-Tertiary of Tasmania. J Paleontol 43:179–182

Wermuth H, Mertens R (1961) Schildkroten, Krokodile, Brückenechsen. G. Fischer, Jena

Wichmann R (1927) Facies lacustres senonianas de los estratos con dinosaurios. Bol Acad Nac Cien 30:386–406

Wieland G (1923) A new Parana Pleurodiran. Am J Sci 5:1–14

Williams EE (1950) Variation and selection in the cervical central articulations of living turtles. Bull Am Mus Nat Hist 94:511–561

Wood R (1971) The fossil Pelomedusidae (Testudines, Pleurodira) of Africa. Unpublished PhD dissertation, Harvard University, Cambridge

Wood R (1976) Two new species of *Chelus* (Testudines: Pleurodira) from the late Tertiary of Northern South America. Breviora 435:1–26

Wood R, Moody R (1976) Unique arrangement of carapace bone in the South American chelid turtle *Hydromedusa maximiliani* (Mikan). Zool J Linn Soc 59:69–78

Woodburne MO, Case JA (1996) Dispersal, Vicariance, and the Late Cretaceous Biogeography from South America to Australia. J Mamm Evo 3:121–161

Zangerl R, Medem F (1958) A new species of chelid turtle *Phrynops (Batrachemys) dalhi* from Colombia. Bull Mus Camp Zool Harv 119:375–390

Chapter 5
North Gondwana Pleurodiran Turtles

Abstract Pelomedusoides comprises five clades (Pelomedusidae, Araripemydidae, Euraxemydidae, Bothremydidae, and Podocnemididae) with a fairly limited distribution in the Southern Hemisphere. With the exception of Pelomedusidae, the clades occurred or occur in South America. According to recent phylogenetic and biogeographic studies, non-chelid eupleurodiran turtles differentiated in the northern region of Gondwana (northeastern South America–northeastern Africa) at least as early as the Early Cretaceous.

Keywords Pelomedusoides • Gondwana • Cretaceous • Cenozoic • Africa • South America

5.1 Pelomedusoides Cope 1868

Gaffney (1972, 1975) was the first to use cladistics to identify synapomorphies for the cryptodiran and pleurodiran clades. Pelomedusidae (sensu lato) and Chelidae, the main groups of crown Pleurodira, have been identified by sharing derived characters in Gaffney's works (e.g., Gaffney 1977; Gaffney and Meylan 1988). Broin (in Antunes and Broin 1988) presented a phylogenetic hypothesis regarding the relationships among Bothremydidae to demonstrate that three different groups (Pelomedusidae, Podocnemididae, and Bothremydidae) are traditionally referred to in the literature as Pelomedusidae (sensu lato). In Antunes and Broin's (1988) paper, this group is elevated to the hyperfamily Pelomedusoides. More recently this category has been used by different authors (i.e., Broin 1988; Meylan 1996; Gaffney and Wood 2002; Gaffney et al. 2006, 2011) to identify non-chelid eupleurodiran turtles.

M.S. de la Fuente et al., *Origin, Evolution and Biogeographic History of South American Turtles*, Springer Earth System Sciences, DOI 10.1007/978-3-319-00518-8_5,

5.1.1 *Podocnemididae* Cope 1868

According to Gaffney et al. (2011), Podocnemididae consists of 20 genera and 30 species. Among them only 3 genera and 8 species are components of the modern chelonian fauna and are distributed in the Southern Hemisphere (Northern South America, and Madagascar) (Fig. 5.1a). This clade extends from the Late Cretaceous to the Recent and occurs in Africa, Asia, Europe, and North and South America. Relationships between extant species based on molecular data and between extant and extinct species based on morphological data have recently been addressed (Vargas-Ramírez et al. 2008; Gaffney et al. 2011, and Cadena 2011, 2012).

Genera Included: *Albertwoodemys* Gaffney, Meylan, Wood, Simons and Almeida Campos 2011, *Bairdemys* Gaffney and Wood 2002, *Bauruemys* Kischlat 1994, *Brontochelys* Gaffney, Meylan, Wood, Simons and Almeida Campos 2011, *Cambaremys* Franca and Langer 2005, *Caninemys* Meylan, Gaffney and Almeida Campos 2009, *Carbonemys* Cadena, Ksepka, Jaramillo and Bloch 2012, *Cerrejonemys* Cadena, Bloch and Jaramillo 2010, *Cordichelys* Gaffney, Meylan, Wood, Simons, Almeida Campos 2011, *Dacquemys* Williams 1954a, *Erymnochelys* Baur 1888, *Lapparentemys* Gaffney, Meylan, Wood, Simons and Almeida Campos 2011, *Latenemys* Gaffney, Meylan, Wood, Simons and Almeida Campos 2011, *Lemurchelys* Gaffney, Meylan, Wood, Simons and Almeida Campos 2011, *Mogharemys* Gaffney, Meylan, Wood, Simons and Almeida Campos 2011, *Neochelys* Broin 1977, *Papulemys* Tong 1998, *Peiropemys* Gaffney, Meylan, Wood, Simons and Almeida Campos 2011, *Peltocephalus* Duméril and Bibron 1835, *Podocnemis* Wagler 1830, *Pricemys* Gaffney, Meylan, Wood, Simons and Almeida Campos 2011, *Roxochelys* Price 1953, *Shweboemys* Swinton 1939, *Stereogenys* Andrews 1901, *Stupendemys* Wood 1976, and *Turkanemys* Wood 2003.

South American Taxa

5.1.1.1 *Peltocephalus* Duméril and Bibron 1835

Type species: *Emys dumeriliana* Schweigger 1812.
Diagnosis: See Gaffney et al. (2011).
Included species: *Peltocephalus dumerilianus* (Schweigger 1812).

Peltocephalus dumerilianus (Schweigger 1812)

Holotype: Muséum National d'Histoire Naturelle Paris, France. Apparently lost.
Diagnosis: As the genus by monotypy (see Gaffney et al. 2011).
Distribution: Orinoco and Amazonas river drainages of northern South America (King and Burke 1989; Iverson 1992) (Fig. 5.1b).
Comments: Williams (1954a, b) identified the specimen MNHN 7893 as *Podocnemis unifilis*, which Fretey (1977) believed was the holotype of *Peltocephalus dumerilianus*, making *Emys trocaxa* the acceptable name. However, after reading Schweigger's description, Pritchard and Trebbau (1984) determined that MNHN 7893 could not be the holotype of *Peltocephalus dumerilianus* because the original description was based on a *Peltocephalus*, not *Podocnemis*.

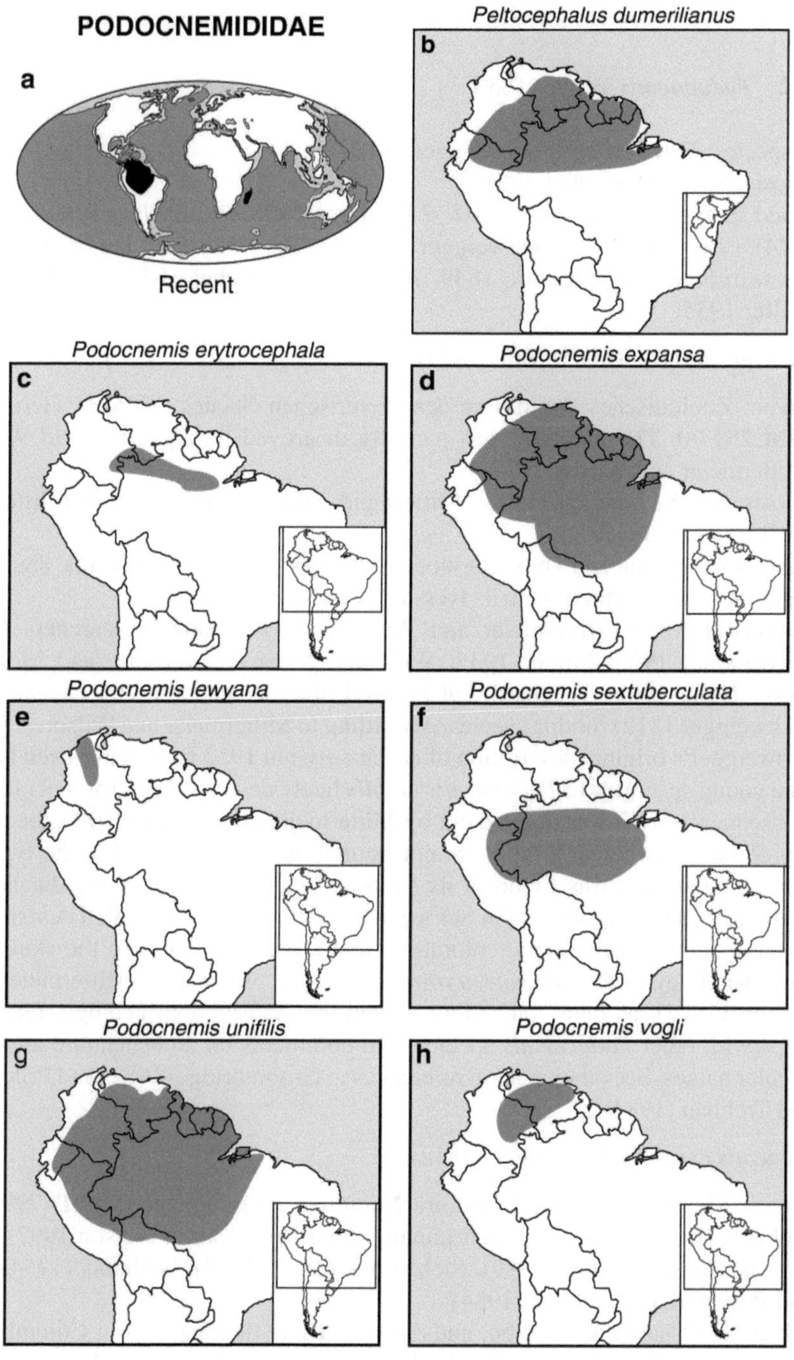

Fig. 5.1 Geographic distribution of Podocnemididae in recent times. (**a**) Global distribution of Podocnemididae. (**b**) Geographic distribution of *Peltocephalus durmerilianus* in South America. (**c**) Geographic distribution of *Podocnemis erythrocephala* in South America. (**d**) Geographic distribution of *Podocnemis expansa* in South America. (**e**) Geographic distribution of *Podocnemis lewyana* in South America. (**f**) Geographic and stratigraphic location of *Podocnemis sextuberculata* in South America. (**g**) Geographic and stratigraphic location of *Podocnemis unifilis* in South America. (**h**) Geographic and stratigraphic location of *Podocnemis vogli* in South America

5.1.1.2 *Podocnemis* Wagler 1830

Type species: *Emys expansa* Schweigger 1812.
Diagnosis: See Gaffney et al. (2011).
Included species: *Podocnemis bassleri* Williams 1956, *Po. erythrocephala* (Spix 1824), *Po. expansa* (Schweigger 1812), *Po. lewyana* Duméril 1852, *Po. sextuberculata* Cornalia 1849, *Po. unifilis* Troschel 1848, and *Po. vogli* Müller 1935.

Podocnemis erythrocephala (Spix 1824)

Holotype: Zoologisches Sammlung des Bayerischen Staates, Münich, Germany. ZSM 2517/0. The holotype was partially destroyed during the World War II (Mittermeier and Wilson 1974).
Diagnosis: See Siebenrock (1902), Mittermeier and Wilson (1974), and Pritchard and Trebbau (1984).
Distribution: Amazon and Upper Orinoco drainages of eastern Colombia, southern Venezuela, and northern Brazil (Iverson 1992) (Fig. 5.1c).
Comments: Prior to Mittermeier and Wilson's (1974) paper, researchers (e.g., Luederwaldt 1926; Müller 1935; Williams 1954a, b; Wermuth and Mertens 1961; Neill 1965; Pritchard 1967) used the designation *Po. cayennensis* (Schweigger 1812) for this species. According to Mittermeier and Wilson (1974), Schweigger's original description of *cayennensis* in 1912 might have been based on a young specimen of *Podocnemis unifilis* lately described by Troschel (1848). Consequently, as it was pointed out by Mittermeier and Wilson (1974), the identity of the Schweigger's types cannot be determined since neither the type of *Emys cayennensis* was found in its supposed repository nor all the diagnostic characters can be recognized in Schweigger's description. A detailed description of the external morphology, coloration, and a brief reference to the skull and shell osteology of *Podocnemis erythrocephala* are provided by Mittermeier and Wilson (1974), as data concerning geographic distribution, ecology, relationships with other *Podocnemis* species, and comments on nomenclature and vernacular names. See other general references in Groombridge (1982) and Pritchard and Trebbau (1984).

Podocnemis expansa (Schweigger 1812)

Syntypes: Muséum National d'Histoire Naturelle, Paris, France. MNHN7997, is probably one of the 3 syntypes originally described (Bour in Iverson 1992).
Diagnosis: See Boulenger (1889), Siebenrock (1902, 1904), Williams (1954a, b), and Pritchard and Trebbau (1984).
Distribution: Orinoco, Essequibo, and Amazon River drainages from Colombia to Bolivia and Brazil (King and Burke 1989; Iverson 1992) (Fig. 5.1d).
Comments: The Arrau side-necked turtle was originally described as *Emys expansa* by Schweigger (1812). Later, Wagler (1830) transferred *expansa* to the genus *Podocnemis*. Anatomical information about this species is presented by Siebenrock (1902), Williams (1954a, b), and Gaffney (1979, 1990). The key to

distinguish *expansa* from remaining *Podocnemis* species is presented by Williams (1954a, b) and Ernst and Barbour (1989). Detailed reviews of the habitat, ecology, population, and conservation status were published by Groombridge (1982) and Pritchard and Trebbau (1984); the latter also presented a morphological description of the largest extant *Podocnemis* (Pritchard and Trebbau 1984).

Podocnemis lewyana Duméril 1852

Syntypes: Muséum National d'Histoire Naturelle de Paris, France. MNHN 8905, MNHN 8360, MNHN 8905 were designed lectotypes by Williams (1954a, b).
Definition: Although neither diagnosis nor definition of this species has been presented, several authors have published comparative information between this and other species (i.e., Boulenger 1889, Siebenrock 1902, and Williams 1954a, b).
Distribution: Restricted to the Magdalena River and Sinu drainages in Colombia (Groombridge 1982; Iverson 1992) (Fig. 5.1e).
Comments: The Magdalena River turtle was named by Duméril in 1852 based on two specimens, one of which (MNHN 8360) was referred to *Podocnemis vogli* by Williams (1954a, b). Anatomical information about this species was provided by Siebenrock (1902) and Williams (1954a, b). A review of the geographic distribution, population status, habitat, ecology, and conservation was presented by Groombridge (1982).

Podocnemis sextuberculata Cornalia 1849

Holotype: Not found.
Diagnosis: There is not a comparative diagnosis for this species. However, a general characterization was published by Ernst and Barbour (1989).
Distribution: Amazon River drainage northern Brazil, southeastern Colombia, and northeastern Peru (Iverson 1992) (Fig. 5.1f).
Comments: The six tubercled Amazon River turtle was named by Cornalia in 1849. Anatomical information about this species was provided by Siebenrock (1902) and Williams (1954a, b). Data on reproductive biology was reported by Vanzolini (1977) and Pritchard (1979). A review of the distribution, population status, habitat, ecology, and conservation of this species is presented by Groombridge (1982).

Podocnemis unifilis Troschel 1848

Holotype: Museum für Naturkunde, Universität Humboldt, Berlin, Germany. ZMB 142. According to King and Burke (1989) and Iverson (1992), this type is lost.
Diagnosis: See Siebenrock (1902), Williams (1954a, b), and Pritchard and Trebbau (1984).
Distribution: Amazon and Orinoco river drainages of Brazil, eastern Colombia, eastern Ecuador, northeastern Peru, northern Bolivia, the Guianas, and Venezuela (Iverson 1992) (Fig. 5.1g).
Comments: Existing and valid name for yellow-spotted river turtle has been discussed by several authors. Among them, Mittermeier and Wilson (1974) and Wermuth and Mertens (1977) have suggested that *Emys cayennensis* Schweigger

would be the oldest name for this species. Likewise, Bour (in Pritchard and Trebbau 1984) suggests that MNHN 8359 is one of the specimens on which Schweigger (1812) based his description of *Emys cayennensis*. To the contrary, Hoogmoed and Grüber (1983) suggested that *E. cayennensis* is a junior synonymous of *Podocnemis expansa* and that *E. dumerilianus* would then be the oldest name available for *Podocnemis unifilis*. Although Iverson (1992) did not dismiss the availability of *Emys cayennensis*, this author suggested using Troschel's (1848) more stable name, *Podocnemis unifilis,* for this species. Available morphological information about *Po. unifilis* is provided by Boulenger (1889), Siebenrock (1902, 1904), Williams (1954a, b), and Pritchard and Trebbau (1984). In particular, Siebenrock (1904) and Pritchard and Trebbau (1984) have presented detailed descriptions of this taxon. Pritchard and Trebbau (1984) have also summarized the data regarding the geographical distribution, habitat, ecology, reproductive biology, and economic importance of *Po. unifilis*.

Podocnemis vogli Müller 1935

Holotype: Zoologisches Sammlung des Bayerischen Staates, Munich, Germany. ZSM, 128/28.

Diagnosis: See Williams (1954a, b) and Pritchard and Trebbau (1984).

Distribution: Restricted distribution to grassland areas of Orinoco drainage of Colombia and Venezuela (Pritchard and Trebbau 1984; Iverson 1992) (Fig. 5.1h).

Comments: The savannah side-necked turtle was named and described by Müller 1935, who has illustrated the skull and views of the carapace and plastron. More recently, Williams (1954a, b) provided a palatal view of the skull, and Alarcon Pardo (1969) has illustrated the head and shell with scales. A detailed morphological description of *Po. vogli* is also presented by Pritchard and Trebbau (1984). Data regarding the ecology and reproductive biology of this species was originally presented by Alarcon Pardo (1969) and Ramo (1982), and a complete synthesis of the biology and conservation of this species was presented by both Groombridge (1982) and Pritchard and Trebbau (1984).

Podocnemis bassleri Williams 1956

Holotype: American Museum of Natural History, New York, USA. AMNH 1662, a skull.

Diagnosis: See Williams (1956).

Locality, horizon, and age: Río Aguaytia, Eastern Peru. Contamana Group, probably Late Miocene (Gaffney et al. 1998; Meylan et al. 2009) (Fig. 5.2a, b).

Comments: This species was named and fully described by Williams (1956) on the basis of a complete skull. Wood and Díaz de Gamero (1971) suggested that *Po. bassleri* is closely related to *Po. expansa*, and Gaffney et al. (2011) noted that the two species are indistinguishable on the basis of the skull, proposing instead that *Po. bassleri* might be synonymous of *Po. expansa*. Disagreement aside, the skull of *Po. bassleri* would be the oldest Cenozoic record of *Podocnemis*.

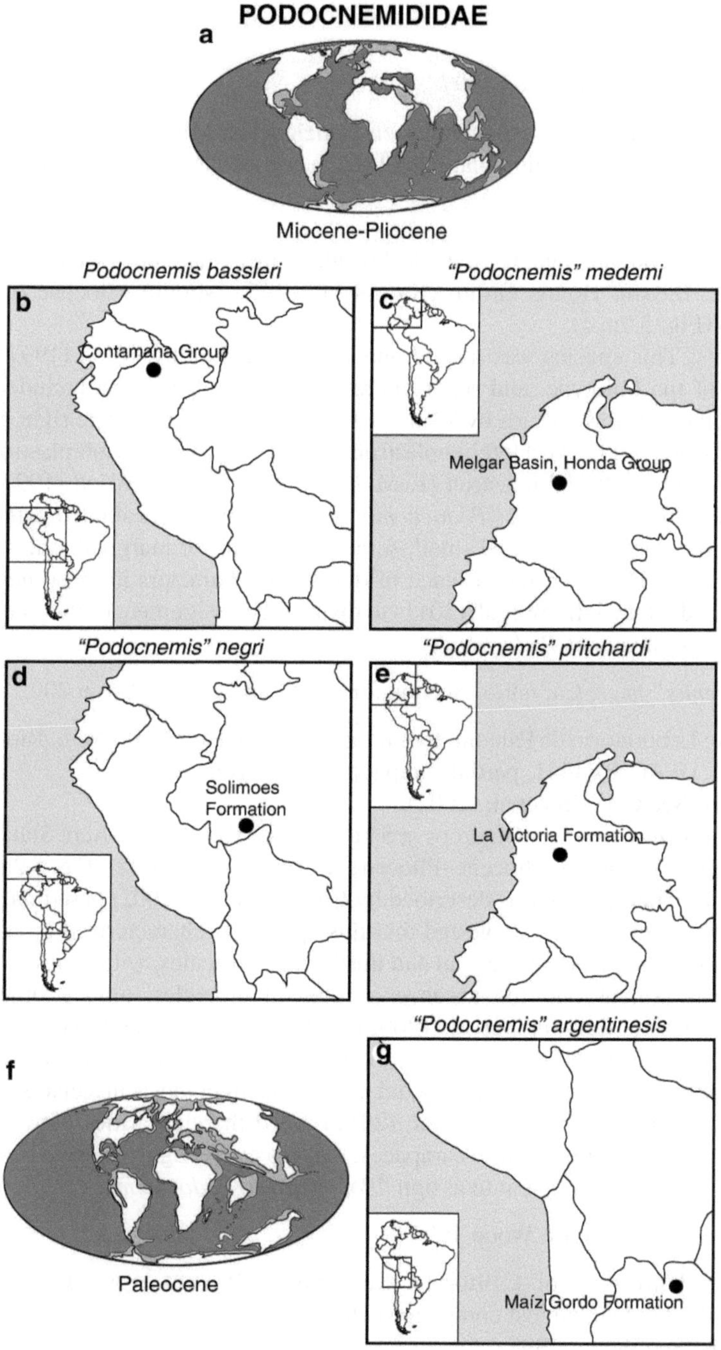

Fig. 5.2 Geographic and stratigraphic distribution of *Podocnemis* (Podocnemididae). (**a**) Paleogeographic reconstruction of continents during the Miocene-Pliocene. (**b**) Geographic and stratigraphic location of *Podocnemis bassleri*. (**c**) Geographic and stratigraphic location of *"Podocnemis" medemi*. (**d**) Geographic and stratigraphic location of *"Podocnemis" negri*. (**e**) Geographic and stratigraphic location of *"Podocnemis" pritchardi*. (**f**) Paleogeographic reconstruction of continents during the Palaeocene. (**g**) Geographic and stratigraphic location of *"Podocnemis" argentinensis*

"*Podocnemis*" *medemi* Wood 1997

Holotype: Instituto de Investigaciones en Geociencias, Minería y Química, Museo Geológico, Bogotá, Colombia. IGM 1863, a partial carapace and an almost complete plastron.
Diagnosis: See Wood (1997).
Locality, horizon, and age: Carmen de Apicalá, upper Magdalena River, Colombia. Melgar Bassin, Honda Group (Kay et al. 1997), Middle Miocene (Guerrero 1997) (Fig. 5.2a, c).
Comments: This species was named and fully described by Wood (1997) on the basis of the holotype, and other, more fragmented specimens included in the hypodigma of this species by Wood (1997), including a complete right hyoplastron (UCMP 125524), a left hyoplastron (IGM 1808), a right xiphiplastron (IGM 1922), and a left xiphiplastron (IGM 2315). According to Wood (1997), "*Po.*" *medemi* differs from other *Podocnemis* species in its size, rounded anterior and posterior plastral lobes, and small, squarish first pair of marginal scales, among other traits. However, the absence of diagnostic characters in *Podocnemis* species conducted Gaffney et al. (2011) to dispute the assignment of this shell-based species to the genus *Podocnemis*.

"Podocnemis" negri Carvalho, Bocquentin, and Lapparent de Broin 2002

Holotype: Laboratorio de Paleontología, Universidade Federal do Acre, Río Branco, Brazil. UFAC-PV4441, partial carapace and plastron.
Diagnosis: See Carvalho et al. (2002).
Locality, horizon, and age: Outcrops at 5 km south of Rio Branco, Acre State, Brazil. Solimões Formation, Miocene–Pliocene (Carvalho et al. 2002) (Fig. 5.2a, d).
Comments: This species was described by Carvalho et al. (2002) based only on the holotype. The authors suggested that this species is characterized by a midline crest extending from the second and third vertebral scales, with an apex between the scale sulcus and the absence of plastral tubercles, among other traits. According to Carvalho et al. (2002), "*Po.*" *negri* is allied with the extant *Podocnemis sextuberculata*. These authors pointed out two characters that supported this classification, but they did not test the hypothesis in a cladistic framework. Even though Gaffney et al. (2011) noted the similarities between "*Po.*" *negri* and *Po. sextuberculata* carapaces, they do not recognize derived characters in *negri*, conducting them to assign "*Po.*" *negri* to *Podocnemis*.

"*Podocnemis*" *pritchardi* Wood 1997

Holotype: University of California, Museum of Paleontology, Berkeley, USA UCMP 63782, an almost complete shell.
Diagnosis: See Wood (1997).
Locality, horizon, and age: Outcrops of the upper Magdalena river valley, Colombia. La Victoria Formation, Miocene (Guerrero 1997; Flynn et al. 1997) (Fig. 5.2a, e).
Comments: This species was named and described by Wood (1997) on the basis of the holotype and additional material included by Wood (1997, p. 156) in the

hypodigma of this species. In agreement with Wood (1997), this species is characterized by an almost rectangular and laterally placed mesoplastra, outer surface of the bridges that curves downward and outward, and very narrow ischial scars situated adjacent to the anal notch. Although Wood used other characters in the diagnosis (e.g., unnatural dorsoventral flattening of the shell, number of neural bones), these have been questioned by Gaffney et al. (2011). Gaffney et al. (2011) accepted the specific entity but dispute the generic assignation.

"*Podocnemis*" *argentinensis* Cattoi and Freiberg 1958

Holotype: Museo Argentino de Ciencias Naturales "Bernardino Rivadavia," Buenos Aires, Argentina. MACN 17988, an almost complete plastron without anterior and posterior ends.

Diagnosis: See Cattoi and Freiberg 1958.

Locality, horizon, and age: Quebrada Queñoal, Humahuaca, Jujuy Province, Argentina. Maíz Gordo Formation (Moreno 1970; Gómez Omil et al. 1989), Upper Palaeocene (Pascual et al. 1978, 1981; Marshall et al. 1983; Gayet et al. 1991) (Fig. 5.2f, g).

Comments: The holotype and a referred specimen (MACN 16553) described and illustrated by Cattoi and Freiberg (1958) were found in different localities (Quebrada Queoñal, the former and Quebrada Ajita, the latter). As it was noted by Broin (1991), this taxon has affinities with *Podocnemis*, although the generic assignment is indeterminable. The plastron does not exhibit the thickening characteristic of *Lapparentemys vilavilensis* (Broin 1971). Furthermore, the gular and intergular scales are short and restricted to the epiplastra, the intergular scale is wide, and the humero-pectoral sulcus crosses over the entoplastron as in *Podocnemis*. Additional shells, smaller than the holotype, have been found at the Río Casa Grande locality in the same lithostratigraphic unit (see Gasparini and Báez 1975; Broin and de la Fuente 1993). These specimens have the plastral morphology similar to that of the holotype, and the anterior margin of the carapace is rounded. Undescribed specimens referable to "*Podocnemis*" *argentinensis* are recognized in the collections of the American Museum of Natural History. However, these specimens are considerably smaller than the holotype. Among them there is an isolated skull that is characterized by a short beak, absence of a groove between the lateral orbits, high maxilla, and week posterior skull elevation. These features suggest a new genus of Podocnemididae (de la Fuente et al. in prep.). More remains recovered from a different area, locality, horizon, and age (northern Guandacol, La Rioja Province, Argentina, Puesto de la Flecha Formation, Oligocene) were also referred to "*Podocnemis*" *argentinensis* (de la Fuente et al. 2003).

5.1.1.3 *Caninemys* Meylan, Gaffney and Almeida Campos 2009

Type species: *Caninemys tridentata* Meylan, Gaffney and Almeida Campos 2009.
Diagnosis: See Meylan, Gaffney and Almeida Campos (2009).
Included species: *Caninemys tridentata* Meylan, Gaffney and Almeida Campos 2009.

Caninemys tridentata Meylan, Gaffney and Almeida Campos 2009

Holotype: Departamento Nacional de Produção Mineral, Divisão de Geología e Minerología, Museu de Paleontología, Rio de Janeiro, Brazil. DNPM-MCT 1496-R, a nearly complete skull.
Diagnosis: See Meylan et al. (2009).
Locality, horizon, and age: Volta de Pedra Pintada, upper Río Juruá, Acre, Brazil. Late Miocene (Meylan et al. 2009) (Fig. 5.3a, b).
Comments: This species was named in 2009 by Meylan, Gaffney, and Almeida Campos on the basis of a skull recovered by L. I. Price in 1962 in the Late Miocene of Acre (Brazil). According to Meylan et al. (2009), *Caninemys tridentata* is characterized by greatly inflated maxillae with a ventral tooth-like process in each one. Although there is no overlap in anatomical elements, Meylan et al. (2009) and Gaffney et al. (2011) suggested the skull-based species *Caninemys tridentata* cannot be distinguished from the shell-based species *Stupendemys geographicus* (Wood 1976). The assignment of the *Caninemys tridentata* skull to *Stupendemys geographicus* cannot be dismissed and is not a chimera as evident in the restoration mounted in the American Museum of Natural History under the direction of Eugene Gaffney (Fig. 6).

5.1.1.4 *Bairdemys* Gaffney and Wood 2002

Type species: *Bairdemys hartsteini* Gaffney and Wood 2002.
Diagnosis: See Gaffney and Wood (2002) and Gaffney et al. (2011).
Included species: *Bairdemys hartsteini* Gaffney and Wood 2002, *Bai. healeyorum* Weems and Knight 2013, *Bai. miocenica* Collins and Lynn 1936, *Bai. sanchezi* Gaffney et al. 2008, *Bai. venezuelensis* Wood and Díaz de Gamero 1971, and *Bai. winklerae* Gaffney et al. 2008.

Bairdemys sanchezi Gaffney, Scheyer, Johnson, Bocquentin and Aguilera 2008

Holotype: Alcaldía del Municipio Urumaco, Colección Rodolfo Sánchez, Venezuela. AMU-CURS 186, skull, mandible, and anterior plastral fragments.
Diagnosis: See Gaffney et al. (2008).
Locality, horizon, and age: Tío Gregorio, Falcón State, Venezuela. Upper Member of the Urumaco Formation, Middle–Late Miocene (Linares 2004; Quiroz and Jaramillo 2010) (Fig. 5.3a, c).
Comments: This species was named and fully described by Gaffney et al. (2008). The excellent preservation of the skull permitted recognizes several autapomorphies. According to Gaffney et al. (2008), the identification of this species is based on the small size, flat triturating surface, and extensive temporal and cheek emarginations.

Bairdemys venezuelensis (Wood and Díaz de Gamero 1971)

Holotype: Laboratorio de Paleontología, Escuela de Geología de la Facultad de Ingeniería de la Universidad Central de Venezuela, Caracas, Venezuela. VF 1176, a complete carapace.

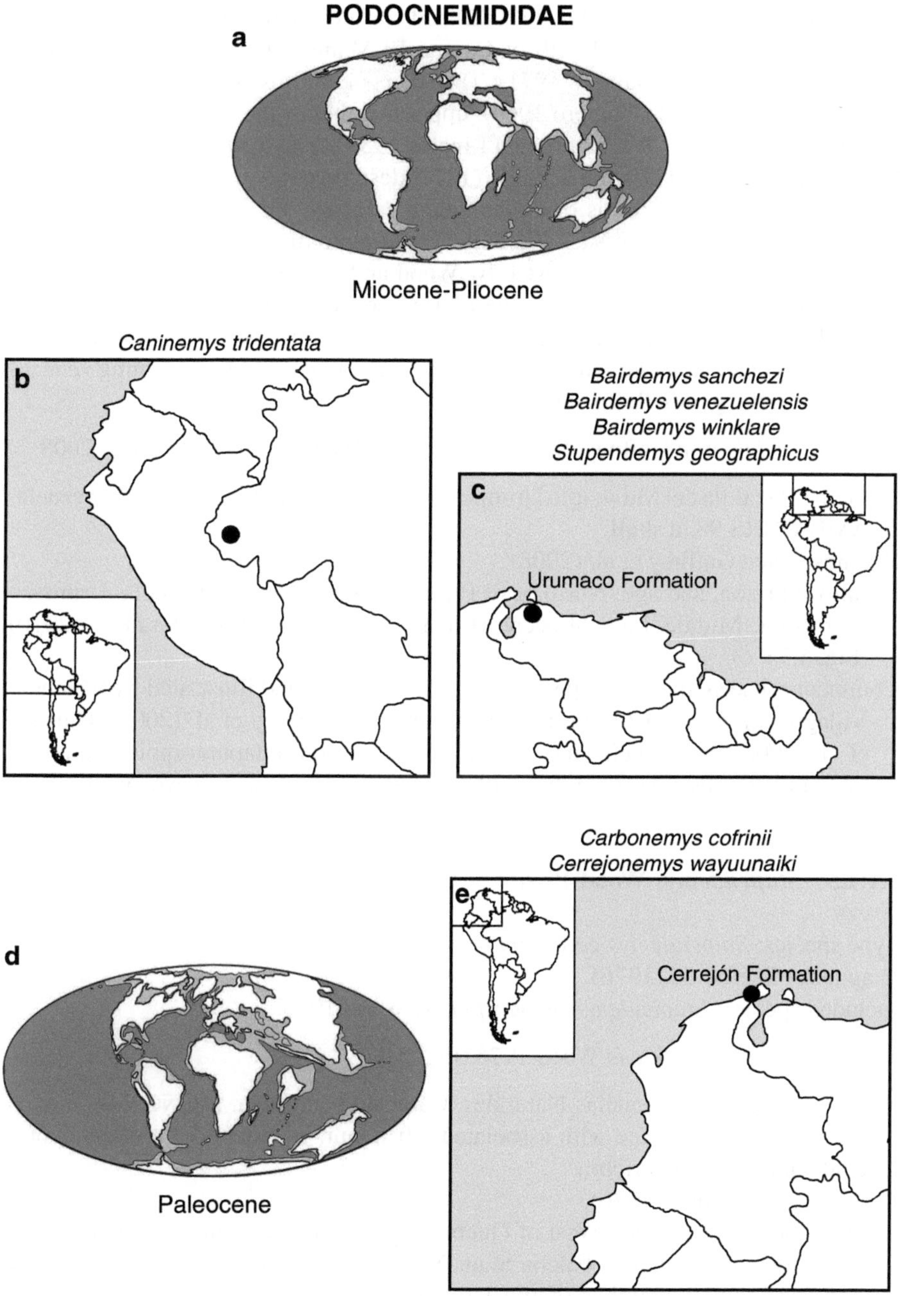

Fig. 5.3 Geographic and stratigraphic distribution of extinct Podocnemididae. (**a**) Paleogeographic reconstruction of continents during the Miocene-Pliocene. (**b**) Geographic location of *Caninemys tridentata* in South America. (**c**) Geographic and stratigraphic location of *Bairdemys sanchezi*, *Bairdemys venezuelensis*, *Bairdemys winklerae*, and *Stupendemys geographicus*. (**d**) Paleogeographic reconstruction of continents during the Palaeocene. (**e**) Geographic and stratigraphic location of *Carbonemys cofrinii* and *Cerrejonemys wayuunaiki*

Diagnosis: See Gaffney and Wood (2002) and Gaffney et al. (2011).

Locality, horizon, and age: North of Campo El Mamón, Falcón State, Venezuela (Wood and Díaz de Gamero 1971). The "capa de huesos" or "capa de tortugas" (Pascual and Díaz de Gamero 1969), upper member of the Urumaco Formation, late Miocene (Díaz de Gamero and Linares 1989) (Fig. 5.3a, c).

Comments: Wood and Díaz de Gamero (1971) described this species as *Podocnemis venezuelensis* on the basis of the complete carapace VF 1176 (holotype) and a fairly complete plastron VF 1173 that, according to the authors, could belong to the same individual. As suggested by Wood and Díaz de Gamero (1971), when the skull was encountered, the shell-based species referred to as *Podocnemis* would be transferred to another genus. In this case, the skull that was analyzed and designated as *venezuelensis* led Gaffney and Wood (2002) assigning *venezuelensis* to the genus *Bairdemys*.

Bairdemys winklerae Gaffney, Scheyer, Johnson, Bocquentin and Aguilera 2008

Holotype: Alcaldía del Municipio Urumaco, Colección Rodolfo Sánchez, Venezuela. AMU-CURS 98, a skull.

Diagnosis: See Gaffney et al. (2008).

Locality, horizon, and age: North of El Picache, Falcón State, Venezuela. Urumaco Formation, Middle–Late Miocene (Linares 2004; Quiroz and Jaramillo 2010) (Fig. 5.3a, c).

Comments: The skull of this species was first described and illustrated by Sánchez-Villagra and Winkler (2006) and later named by Gaffney et al. (2008). Gaffney et al. (2008) based their characterization on several autapomorphic characters including the (large) size of the skull, elongated snout, and concave labial ridges.

5.1.1.5 *Stupendemys* Wood 1976

Type species: *Stupendemys geographicus* Wood 1976.

Diagnosis: See Wood (1976).

Included species: *Stupendemys geographicus* Wood 1976.

Stupendemys geographicus Wood 1976 (Fig. 5.4)

Holotype: Museo de Ciencias Naturales, Caracas, Venezuela. MCNC 244, medial portion of the carapace with associated left femur, fragments of a scapulocoracoid, and a caudal vertebra.

Diagnosis: See Wood (1976).

Locality, horizon, and age: West of Quebrado Gregorio, El Mamón oil field, north of the town of Urumaco, Falcón State, Venezuela."Capa de huesos" or "capa de Tortugas," upper member of the Urumaco Formation, Middle–Late Miocene (Linares 2004; Quiroz and Jaramillo 2010) (Fig. 5.3a, c).

Comments: S*tupendemys geographicus,* the largest turtle ever described, was named by Wood (1976) on the basis of the holotype and associated materials, among which there is a well-preserved shell housed at the Museum of

Fig. 5.4 Life reconstruction of *Stupendemys geographicus* from the Miocene of the Urumaco Formation, Venezuela

Comparative Zoology (MCZ P 4376) with a midline carapace length of 218 cm. Recently Aguilera (2004) fide Sánchez-Villagra and Scheyer (2010) have reported and additional *Stupendemys* specimen from Urumaco with a carapace length of 310 cm. Aside from the enormity of this species, it was characterized (among other traits mentioned in Wood 1976) by a thickened and upturned nuchal anterior margin of the carapace, massive cervical vertebrae, a very massive humerus with expanded proximal and distal ends a triangular shaft cross section, and a massive femur that is greatly flattened dorsoventrally. Gaffney et al. (1998) have also cited the peculiar morphology of the dorsal scapular process (strongly bowed) of *Stupendemys geographicus*. Additional specimens of *Stupendemys* from southwestern Amazonia were described by Lapparent de Broin et al. (1993), Bocquentin and Negri (1993), and Gaffney et al. (1998). Lapparent de Broin et al. (1993) attributed to *Stupendemys* sp. a left second costal bone, a first left peripheral, and a right humerus from Mio-Pliocene outcrops from Patos locality, Alto Río Acre, Brazil (see Frailey 1986; Lapparent de Broin et al. 1993). These authors suggested, based on the morphology of this peripheral bone, the presence of a deep midline notch and collar-like structure at the anterior margin of the carapace that is equivalent to the structure shown in the holotype of *Stupendemys geographicus*. Later, Bocquentin and Melo (2006) assigned three elements described by Lapparent de Broin et al. (1993) to the new species *Stupendemys souzai* as well as two vertebrae described by Bocquentin and Negri (1993) and six other bone elements. According to Meylan et al. (2009), the assignation of these isolated skeletal bones to the genus *Stupendemys* is not well justified by Bocquentin and Melo (2006). Likewise, Meylan et al. (2009) suggested that the 11 elements included by Bocquentin and Melo (2006) in the hypodigm of *Stupendemys souzai* may belong to more than one individual, and some of them may not even belong to

Stupendemys. Scheyer and Sánchez-Villagra (2007) studied the carapace bone histology of this species. Additional information regarding different interpretations of *Stupendemys geographicus* lifestyle is summarized in Sánchez-Villagra and Scheyer (2010).

5.1.1.6 *Carbonemys* Cadena, Ksepka, Jaramillo and Bloch 2012

Type species: *Carbonemys cofrinii* Cadena, Ksepka, Jaramillo and Bloch 2012.
Diagnosis: See Cadena et al. (2012).
Species included: Only *Carbonemys cofrinii* Cadena, Ksepka, Jaramillo and Bloch 2012.

Carbonemys cofrinii Cadena, Ksepka, Jaramillo and Bloch 2012

Holotype: According to Cadena et al. (2012), the holotype of this species is housed in two different institutions, in the Vertebrate Paleontology Florida Museum of Natural History, University of Florida, Gainesville, USA, and in the Instituto Colombiano de Investigaciones Mineras INGEOMINAS, Bogotá, Colombia. This particularity is not explained by Cadena et al. (2012). UF/IGM 41, nearly complete skull.
Diagnosis: See Cadena et al. (2012).
Locality, horizon, and age: La Puente Pit, Cerrejón Coal Mine, Guajira Peninsula, Colombia. Middle part of the Cerrejón Formation, Late Palaeocene (58–55 Ma), palynological zone Cu-02 (Jaramillo et al. 2007) (Fig. 5.3d, e).
Comments: Different phylogenetic analyses designed to test the phylogenetic position of the podocnemidid *Carbonemys cofrinii* were described by Cadena et al. (2012). A morphological-based analysis identified *Carbonemys cofrinii* as a member of Erymnochelyinae. According to Cadena et al.'s (2012) arrangement, the clade Erymnochelyinae includes two subclades, one of which contains the extant species *Erymnochelys madagascariensis* and *Peltocephalus dumerilianus* and the other uniting *Carbonemys cofrinii* with *Dacquemys paleomorpha*, *Shweboemys antiqua*, *Stereogenys cromeri*, and *Bairdemys* spp. The combined molecular–morphological analysis identified *Peltocephalus dumerilianus* and *Erymnochelys madagascariensis* as successive sister taxa to Podocnemidinae, and confirmed the results of previous molecular studies (Noonan and Chippindale 2006; Vargas-Ramírez et al. 2008).

5.1.1.7 *Cerrejonemys* Cadena, Bloch and Jaramillo 2010

Type species: *Cerrejonemys wayuunaiki* Cadena, Bloch, and Jaramillo 2010.
Diagnosis: See Cadena et al. (2010).
Species included: *Cerrejonemys wayuunaiki* Cadena, Bloch, and Jaramillo 2010.

Cerrejonemys wayuunaiki Cadena, Bloch and Jaramillo 2010

Holotype: According to Cadena et al. (2010), the holotype of this species is housed in two different institutions: in the Florida Museum of Natural History, Vertebrate Paleontology Collections, Gainesville, Florida, USA, and in the Museo Geológico at the Instituto Nacional de Investigaciones en Geociencias, Bogotá, Colombia. This peculiar situation is not explained in Cadena et al.'s (2010) paper. UF/IGM 33, skull and shell.

Diagnosis: See Cadena et al. (2010).

Locality, horizon, and age: Cerrejón Coal Mine, Guajira Peninsula, Colombia. Cerrejón Formation, Middle–Upper Palaeocene (Cadena et al. 2010) (Fig. 5.3d, e).

Comments: The phylogenetic analysis conducted by Cadena et al. (2010) suggests that *Cerrejonemys* is the sister taxon of *Podocnemis* on the basis of a small postorbital and the jugal–parietal contact. However, because of the ambiguity in the sutures on the crushed holotype, Gaffney et al. (2011) proposed designating this species as Podocnemididae incertae sedis. These authors noted that *Cerrejonemys* has a podocnemidid cavum pterygoid but is also characterized by the absence of the *Podocnemis*-like triturating surface ridges.

5.1.1.8 *Lapparentemys* Gaffney, Meylan, Wood, Simons, Almeida Campos 2011

Type species: ?*Roxochelys vilavilensis* Broin 1971.

Diagnosis: See Broin (1971) and Gaffney et al. (2011).

Species included: *Lapparentemys vilavilensis* (Broin 1971).

Lapparentemys vilavilensis (Broin 1971)

Holotype: Muséum National d'Histoire Naturelle, Paris, France. MNHNP VIL 1, a complete carapace and plastron.

Diagnosis: See Broin (1971) and Gaffney et al. (2011).

Locality, horizon, and age: 2 km south of Vilavila (= Villa Vizcarra), Cochabamba Province, Bolivia. Santa Lucía Formation, Early Palaeocene (Broin 1991; Gayet et al. 1991) (Fig. 5.5a, b).

Comments: The species *vilavilensis* was described and named by France de Broin in 1971 on the basis of the holotype (Broin 1971, plate 31) and other materials such as a lower jaw and remains of another skull (Broin 1971, plate 32). This species was referred tentatively to the genus *Roxochelys* by Price 1953. More recently, Broin (1991) in a general account about the Bolivian fossil turtle fauna provided additional descriptions and illustrated this taxon (Broin 1991, plate 1–2). Broin (1991) recognized the polymorphic nature of this species. However, the most recent reviewers of this species (Gaffney et al. 2011) suggested that more than one taxon could be included in the hypodigm of *vilavilensis* and, consequently, they restricted this species to the holotype. Gaffney et al.

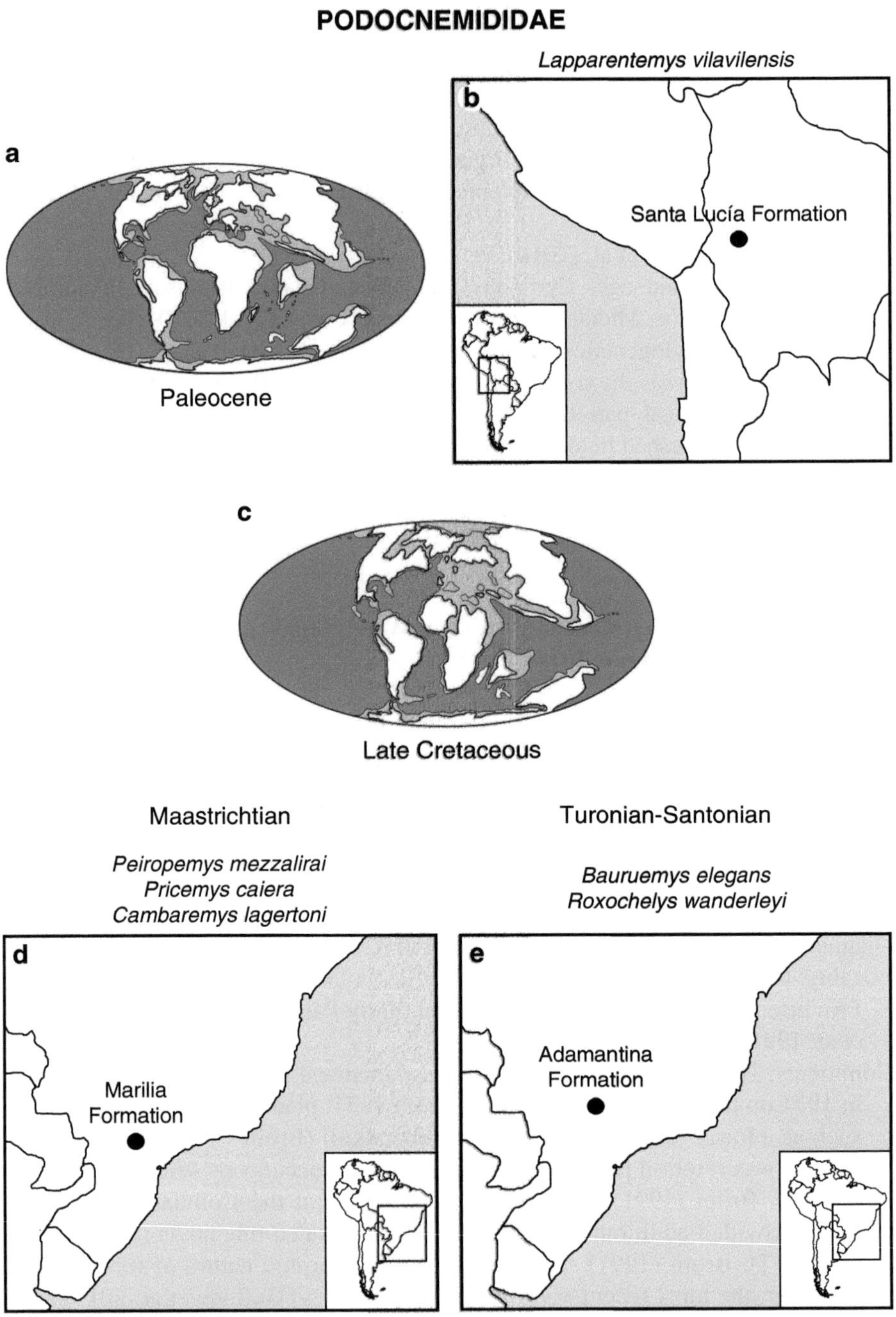

Fig. 5.5 Geographic and stratigraphic distribution of extinct Podocnemididae. (**a**) Paleogeographic reconstruction of continents during the Palaeocene. (**b**) Geographic and stratigraphic location of *Lapparentemys vilavilensis*. (**c**) Paleogeographic reconstruction of continents during the Late Cretaceous. (**d**) Geographic and stratigraphic location of the Maastrichtian species *Peiropemys mezzalirai*, *Pricemys caiera*, and *Cambaremys lagertoni*. (**e**) Geographic and stratigraphic location of the Turonian–Santonian species *Bauruemys elegans* and *Roxochelys wanderleyi*

(2011) also removed *vilavilensis* from genus *Roxochelys* and created a new genus (*Lapparentemys*) for this species on the basis of the redescription of new specimens of *Roxochelys wanderleyi* (type species of genus *Roxochelys*).

5.1.1.9 *Peiropemys* Gaffney, Meylan, Wood, Simons, Almeida Campos 2011

Type species: *Peiropemys mezzalirai* Gaffney, Meylan, Wood, Simons, and Almeida Campos 2011.
Diagnosis: See Gaffney, Meylan, Wood, Simons, and Almeida Campos (2011).
Species included.: *Peiropemys mezzalirai* Gaffney, Meylan, Wood, Simons, and Almeida Campos 2011.

Peiropemys mezzalirai Gaffney, Meylan, Wood, Simons, and Almeida Campos 2011

Holotype: Divisão de Geologia e Mineralogia, Departamento Nacional de Produção Mineral, Rio de Janeiro, Brazil. MCT 1497-R, an almost complete skull.
Diagnosis: See Gaffney, Meylan, Wood, Simons, and Almeida Campos (2011).
Locality, horizon, and age: Caiera Quarry (Campos and Kellner 1999; Kellner et al. 2005) at the southern end of Serra do Vedinha, near the old Peirópolis railroad station, near the city of Uberaba, Minas Gerais, Brazil. Sierra de Galga Member, Marília Formation (Candeiro 2009 and references therein), Upper Cretaceous (Fig. 5.5c, d).
Comments: *Peiropemys mezzalirai* was named and described by Gaffney et al. (2011) on the basis of a skull. Although there is no referred material, Gaffney et al. (2011) have suggested that the shell-based taxon described informally as Peiropolis B could be referred to *Peiropemys*. According to Gaffney et al.'s (2011) parsimony analysis, *Peiropemys* is nested as the sister taxon of *Pricemys* plus *Lapparentemys*. This clade is the sister taxon of remaining Podocnemididae, with the exception of *Bauremys*.

5.1.1.10 *Pricemys* Gaffney, Meylan, Wood, Simons, Almeida Campos 2011

Type species: *Pricemys* Gaffney, Meylan, Wood, Simons, and Almeida Campos 2011).
Diagnosis: See Gaffney, Meylan, Wood, Simons, and Almeida Campos (2011).
Species included: Only *Pricemys caiera* Gaffney, Meylan, Wood, Simons, and Almeida Campos 2011.

Pricemys caiera Gaffney, Meylan, Wood, Simons, Almeida Campos 2011

Holotype: Divisão de Geologia e Mineralogia, Departamento Nacional de Produção Mineral, Rio de Janeiro, Brazil. MCT 1498-R, a partial skull that preserves a braincase articulated with a left quadrate–squamosal and an isolated right maxilla–jugal.
Diagnosis: See Gaffney, Meylan, Wood, Simons, and Almeida Campos (2011).
Locality, horizon, and age: Caiera Quarry, near village of Peirópolis, near Uberaba, Mina Gerais State, Brazil. Serra da Galga Member, Marília Formation, Upper Cretaceous (Candeiro 2009 and references therein) (Fig. 5.5c, d).

Comments: This is another species recently named by Gaffney et al. (2011). *Pricemys caiera* is very closely related to *Peiropemys mezzalirai*, and the main skull differences (e.g., width at condylus mandibularis, width percentage, triturating surface, height at condylus occipitalis, height ratio, triturating surface, condylus occipitalis, antrum postoticum, fossa precolumellaris, and scale arrange on parietals) between the species are summarized by Gaffney et al. (2011, Table 2).

5.1.1.11 *Cambaremys* Franca and Langer 2005

Type species: *Cambaremys langertoni* Franca and Langer 2005.
Diagnosis: See Franca and Langer (2005).
Species included: *Cambaremys langertoni* Franca and Langer 2005.

Cambaremys langertoni Franca and Langer 2005

Holotype: Centro de Pesquisas Paleontológicas "Llewellyn Ivor Price," Uberaba, Brazil. CCP 0252, a partial carapace and a small portion of the plastron, limbs, and vertebral elements (Franca and Langer 2005).
Diagnosis: See Franca and Langer (2005).
Locality, horizon, and age: Serra do Veadinho Area, Municipalidad de Uberaba, proximity of Peirópolis, Minas Gerais, Brazil. Serra do Galga Member, Marília Formation (Late Cretaceous), probably Maastrichtian (Franca and Langer 2005) (Fig. 5.5c, d).
Comments: This species is only known through the holotype, first described by Franca (2004, in an unpublished dissertation) and later described and named by Franca and Langer (2005). According to Gaffney et al. (2011), this taxon is characterized by wide, hexagonal second, third, and fourth vertebrals; a long and narrow nuchal bone; axillary buttresses that do not extend on the second costal bone but reach the second peripheral; and unguttered bridge peripherals, among other traits. There is a general agreement (Broin 1991; Franca and Langer 2005; Gaffney et al. 2011) that several taxa are present among the specimens recovered from Late Cretaceous horizon in the proximity of Peiropolis. Among them, Gaffney et al. (2011) identified three shell morphotypes (*Cambarenys langertoni*, Peiropolis A, and Peiropolis B) and two different skulls (*Pricemys caiera* and *Peiropemys mezzalirai*). Although Romano et al. (2009, 2013) suggested that *Cambaremys langertoni* could be a juvenile of *Roxochelys wanderleyi*, Gaffney et al. (2011) recognized autapomorphic characters in the holotype of *Cambaremys* to consider it a distinct taxonomic entity.

5.1.1.12 *Bauruemys* Kischlat 1994

Type species: *Podocnemis elegans* Suárez 1969.
Diagnosis: See Suárez (1969) and Gaffney et al. (2011).
Species included: Only *Bauruemys elegans* (Pacheco 1969).

Fig. 5.6 Life reconstruction of *Bauruemys elegans* from the Late Cretaceous of the Adamantina Formation, Brazil

Bauruemys elegans (Pacheco 1969) (Fig. 5.6)

Holotype: Divisão de Geología e Mineralogia, Departamento Nacional de Produção Mineral, Rio de Janeiro, Brazil. MCT 1942-R, skull and shell with scapulocoracoids, humerus, and femur (see discussion in Oliveira and Romano 2007; Gaffney et al. 2011).

Diagnosis: See Suárez (1969) and Gaffney et al. (2011).

Locality, horizon, and age: "Tartarugito" locality near kilometer 736 of the old Sorocabana Railway branch, Pirapozinho, São Paulo, State, Brazil (Bertini et al. 2006; Romano and Azevedo 2007; Oliveira and Romano 2007). Adamantina Formation, Upper Cretaceous (Suárez 2002; Bertini et al. 2006) (Fig. 5.5c, e).

Comments: This species was named *Podocnemis elegans* by Suárez (1969). More recently, Kischlat (1994) proposed to remove this species from the genus *Podocnemis* and designate a new genus named *Bauruemys*. Several authors (e.g., Kischlat 1994, 1996; Romano and Azevedo 2006; Matiazzi 2007; Gaffney et al. 2011) have proposed a hypothesis about the phylogenetic relationships of *Bauruemys*. For example, Romano and Azevedo (2006) have suggested that *Bauruemys* is the sister taxa of Podocnemididae, while Gaffney et al. (2011) included this species in the podocnemidid clade Bauruemydinae as the sister group of the remaining Podocnemididae.[1]

5.1.1.13 *Roxochelys* Price 1953

Type species: *Roxochelys wanderleyi* Price 1953.
Diagnosis: See Gaffney et al. (2011).
Included species: *Roxochelys wanderleyi* Price 1953.

[1]Courtesy of the American Museum of Natural History.

Roxochelys wanderleyi Price 1953

Holotype: Divisão de Geologia e Mineralogia, Departamento Nacional de Produção Mineral, Rio de Janeiro, Brazil. MCT-216R, a partial carapace and an anterior plastral lobe.

Diagnosis: See Gaffney et al. (2011).

Locality, horizon, and age: Aracatuba, Jubia Municipality of Mirandópolis, São Paulo State, Brazil. Adamantina Formation, Turonian–Santonian (Price 1953; Candeiro et al. 2006) (Fig. 5.5c, e).

Comments: After a detailed description of the DNPM specimen (a partial carapace and plastron) was used by Staesche (1937) to erect *Podocnemis brasiliensis*, Price (1953) concluded that this material belongs to two different specimens. Llewellyn Price named and described *Roxochelys wanderleyi* on the basis of Staesche's *Podocnemis brasiliensis* carapace and an additional anterior plastral lobe (Price 1953, Figs. 3–4, and plates 1–2) and applied this binominal designation to the associated plastron (Price 1953, Figs. 1–2). An additional specimen of *Roxochelys wanderleyi* (a complete carapace) is illustrated by Romano et al. (2009, 2013). Although a synonymy between *Roxochelys wanderleyi* and "*Podocnemis*" *harrisi* was proposed by Broin (1991), on the basis of the shell ornamentation, this suggestion has been dismissed by other researchers. Finally, according to Oliveira and Romano (2007) and Gaffney et al. (2011), "*Podocnemis*" *harrisi* and "*Podocnemis*" *brasiliensis* could be considered *nomina dubia*.

5.1.2 Basal Podocnemidinura Cope 1868

5.1.2.1 *Portezueloemys* de la Fuente 2003

Type species: *Portezueloemys patagonica* de la Fuente 2003.

Diagnosis: See de la Fuente (2003).

Species included: Only *Portezueloemys patagonica* de la Fuente 2003.

Portezueloemys patagonica de la Fuente 2003 (Fig. 5.7)

Holotype: "Carmen Funes" Museum of Plaza Huincul, Neuquén Province, Argentina. MCF-PVPH-338, a skull and partial carapace and plastron.

Diagnosis: See de la Fuente (2003) and de la Fuente (2007).

Locality, horizon, and age: Sierra del Portezuelo, Neuquén Province, Argentina. Portezuelo Formation (see Leanza 1999; Leanza et al. 2004), Upper Cretaceous (Late Turonian–Early Coniacian) (Fig. 5.8a, b).

Comments: *Portezueloemys patagonica* is a member of Pleurodira Pelomedusoides (sensu Broin 1988) based on the reduction of the vomer to the anterior interchoanal part and rounded lateral mesoplastra. Other derived characters including a

Fig. 5.7 Life reconstruction of *Portezueloemys patagonica* from the Late Cretaceous of the Portezuelo Formation, Argentina

cavum pterygoidei formed by the basisphenoid, pterygoid, prootic, and quadrate underlain by the pterygoid and basisphenoid permit referral of this taxon to Podocnemidinura.[2]

According to de la Fuente (2003), although the skull and shell morphology of *Portezueloemys patagonica* is podocnemid-like in the general arrangement of bones and scales, the pterygoid wings stop at the border of the infratemporal fossa and do not extend posteromedially up to the suture of the basisphenoid as in Podocnemididae. The skull, as in *Podocnemis*, is rather flat and oblong, with small orbits directed dorsoventrally and with a narrow interorbital space, but it retains some plesiomorphic characters such as a foramen jugular posterious confluent with the fenestra postotica and a well-developed paraoccipital process. The shell is oval in shape, rather than quadrangular, and moderately high with a rounded anterior border. Several phylogenetic analyses (de la Fuente 2003; Romano and Azevedo 2006; Gaffney et al. 2011; Cadena et al. 2012) have recovered *Portezueloemys* as a stem Podocnemididae.

5.1.2.2 *Brasilemys* Lapparent de Broin 2000

Type Species: *Brasilemys josai* Lapparent de Broin 2000.
Diagnosis: See Lapparent de Broin (2000) and Gaffney et al. (2006).
Species included: *Brasilemys josai* Lapparent de Broin 2000.

[2]Courtesy of Indiana University Press.

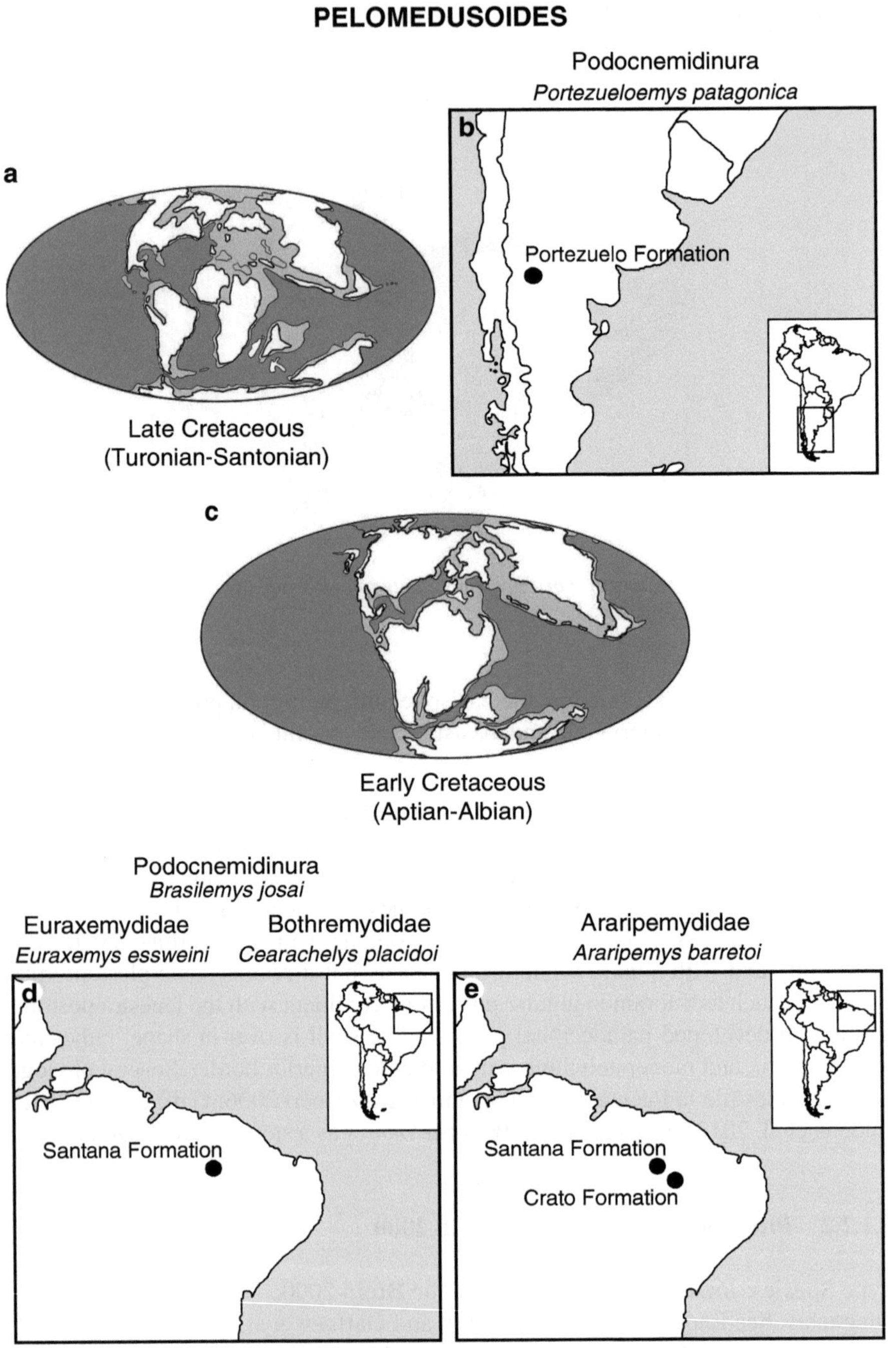

Fig. 5.8 Geographic and stratigraphic distribution of extinct Pelomedusoides. (**a**) Paleogeographic reconstruction of continents during the Late Cretaceous. (**b**) Geographic distribution of the podocnemidinuran *Portezueloemys patagonica* in South America. (**c**) Paleogeographic reconstruction of continents during the Early Cretaceous. (**d**) Geographic and stratigraphic location of the podocnemidinuran *Brasilemys josai*, the euraxemydid *Euraxemys essweini*, and the bothremydid *Cearachelys placidoi*. (**e**) Geographic and stratigraphic location of the araripemydid *Araripemys barretoi*

Brasilemys josai Lapparent de Broin 2000

Holotype: Museu de Geologia de Barcelona, Barcelona, Spain. MGB, 37911 skull and shell of one specimen (Lapparent de Broin 2000).
Diagnosis: See Lapparent de Broin (2000) and Gaffney et al. (2006).
Locality, horizon, and age: Chapada do Araripe, Ceará State, Brazil (Lapparent de Broin 2000). Romualdo Member, Santana Formation, Albian (Lapparent de Broin 2000) (Fig. 5.8c, d).
Comments: *Brasilemys josai* was named and described in detail by Lapparent de Broin (2000) on the basis of the holotype. The branching diagram presented in Lapparent de Broin (2000) and the computer-assisted cladogram in de la Fuente (2003) show *Brasilemys* as the sister taxon of *Hamadachelys* plus Podocnemididae.

5.1.3 Bothremydidae Baur 1891

The clade Bothremydidae is a large and diverse clade of pelomedusoid turtles of Pangaean distribution, extending from the Albian to the Eocene in North and South America, Europe, Africa, and India. According to Gaffney et al.'s (2006) analysis, Bothremydidae is a monophyletic clade, a designation supported by the presence of a wide exoccipital–quadrate contact, a Eustachian tube separated, from the incisura columella auris (usually by bone) to form a bony canal for the stapes, the absence of a fossa precolumellaris, a supraoccipital-quadrate contact (except in the tribe Taphrosphyini), and a posterior enlargement of the fossa orbitalis.

Genera included: *Araiochelys* Gaffney, Tong and Meylan 2006, *Arenila* Lapparent de Broin and Werner 1998, *Azzabaremys* Gaffney, Moody and Walker 2001c, *Cearachelys* Gaffney, Tong, and Hirayama 2001, *Chedighaii* Gaffney, Tong and Meylan 2006, *Elochelys* Nopcsa 1931, *Foxemys* Tong, Gaffney and Buffetaut 1998, *Galianemys* Gaffney, Tong and Meylan 2002, *Kurmademys* Gaffney, Chatterjee and Rudra 2001, *Labrastochelys* Gaffney, Tong and Meylan 2006, *Nigeremys* Broin 1977, *Phosphatochelys* Gaffney and Tong 2003, *Polysternon* Portis 1882, *Rhothonemys* Gaffney, Tong and Meylan 2006, *Rosasia* Carrington da Costa 1940, *Sankuchemys* Gaffney, Sahni, Schleich, Singh and Srivastava 2003, *Taphrosphys* Cope 1869, *Ummulisani* Gaffney and Tong 2003, and *Zolhafah* Lapparent de Broin and Werner 1998.

South American Taxa

5.1.3.1 *Cearachelys* Gaffney, Campos, and Hirayama 2001b

Type species: *Cearachelys placidoi* Gaffney, Campos, and Hirayama 2001b.
Diagnosis: See Gaffney et al. (2001a) and Gaffney et al. (2006).
Species included: *Cearachelys placidoi* Gaffney, Campos, and Hirayama 2001b

Fig. 5.9 Life reconstruction of *Cearachelys placidoi* from the Lower Cretaceous of the Santana Formation, Brazil

Cearachelys placidoi Gaffney, Campos, and Hirayama 2001b (Fig. 5.9).

Holotype: Museu Paleontologico de Santana do Cariri, Ceara State, Brazil. MPSC (without number), a partial skull, shell cervical, and limbs.

Diagnosis: See Gaffney et al. (2001a, b, c) and Gaffney et al. (2006).

Locality, horizon, and age: Santana do Cariri, Ceará, Brazil, ? Romualdo Member of the Santana Formation, Albian in age (Maisey 1990, 1991) (Fig. 5.8c, d).

Comments: This South American bothremydid species was erected by Gaffney et al. (2001a, b, c) on the basis of the holotype and one referred specimen. The authors referred this new taxon to Bothremydidae and concluded that it shows affinities with *Bothremys*, *Rosasia*, *Foxemys*, and *Zolhafah*. More recently, Gaffney et al. (2006) presented a detailed redescription of *Cereachelys placidoi* and included it in a parsimony analysis. In their analysis, Gaffney et al. (2006) grouped *Cea. placidoi* with *Galianemys* species (*G. emringeri* and *G. whitei*) to form the clade Cearachelyinae.

5.1.4 Euraxemydidae Gaffney, Tong and Meylan 2006

Euraxemydidae was recently named by Gaffney et al. (2006) and includes two taxa: *Euraxemys essweini*, from the Albian Santana Formation of Brazil, and *Dirqadim schaefferi*, from the Cenomanian Kem Kem beds of Morocco. Gaffney et al. (2006) reported that the taxa in this group share the following synapomorphies: possession of a medial process of the quadrate, which partially covers the prootic, and a narrow contact with the ventral process of the exoccipital. Euraxemydidae is hypothesized by Gaffney et al. (2006) to be the sister group of Podocnemidoidea.

Genera included: *Euraxemys* Gaffney, Tong and Meylan 2006 and *Dirqadim* Gaffney, Tong and Meylan 2006.

5.1.4.1 *Euraxemys* Gaffney, Tong and Meylan 2006

Type species: *Euraxemys essweini* Gaffney, Tong and Meylan 2006.
Diagnosis: See Gaffney et al. (2006).
Species included: *Euraxemys essweini* Gaffney, Tong and Meylan 2006.

Euraxemys essweini Gaffney, Tong and Meylan 2006

Holotype: Forschungsinstitute Senckenberg, Frankfurt, Germany. FR 4922, a nearly complete skeleton.
Diagnosis: See Gaffney et al. (2006).
Locality, horizon, and age: Araripe basin, Ceara, Brazil. Santana Formation, Albian Early Cretaceous (Maisey 1990, 1991) (Fig. 5.8c, d).
Comments: Initially, Gaffney and Meylan (1991) referred *Euraxemys essweini* as a "primitive pelomedusid turtle." More recently, Meylan (1996) also concluded that *Euraxemys* (FR 4922 at that time) is a member of Araripemydidae. Contrarily to Meylan's 1996 hypothesis, Gaffney et al. (2006) proposed that Euraxemydidae (*Euraxemys*+*Dirqadim*) is related to Podocnemidoidea. The skull, lower jaw, and shell morphology was described in detail by Gaffney et al. (2006).[3]

5.1.5 Araripemydidae Price 1973

Araripemydidae consists of two taxa, *Araripemys barretoi* Price 1973 from the Aptian–Albian of Brazil and *Laganemys tenerensis* Sereno and Elshafie 2013 from the mid-Cretaceous of Niger. Although Broin (1980) also assigns *Taquetochelys decorata* Broin 1980 from the Aptian of Niger to this clade, Gaffney et al. (2006) consider this species as Pelomedusoides incertae sedis. Araripemydidae is characterized by the presence of very thin skull and narrow maxillar and lower jaw triturating surfaces, a long neck, and a flat and textured shell. Other shell characteristics (e.g., absence of mesoplastra and first costals that either reach the margin of the shell or do not) are differential traits between African and South American taxa (Meylan 1996; Gaffney et al. 2006; Sereno and Elshafie 2013). In agreement with the description of new material of *Araripemys barretoi* and the recent parsimony analysis, Gaffney et al. (2006) suggested that Araripemydidae is either the sister group to all other Pelomedusoides or the sister group to Pelomedusidae, but these relationships are only weakly supported in their analysis. More recently, a new taxon of Araripemydidae (*Laganemys tenerensis*) from Illumeden Basin, Elrhaz Formation (Niger), was described by Sereno and ElShafie (2013). These new long-necked Pelomedusoides, which are similar in morphology to *Araripemys barretoi*, provide new information concerning faunal interchange between Africa and South America in the mid-Cretaceous.

Genera included: *Araripemys* Price 1973, *Laganemys* Sereno and ElShafie 2013.

---

[3]Courtesy of the American Museum of Natural History.

Fig. 5.10 Life reconstruction of *Araripemys barretoi* from the Lower Cretaceous of the Santana Formation, Brazil

5.1.5.1 *Araripemys* Price 1973

Type species: *Araripemys barretoi* Price 1973.
Diagnosis: See Meylan (1996) and Gaffney et al. (2006).
Species included: *Araripemys barretoi* Price 1973.

Araripemys barretoi Price 1973 (Fig. 5.10)

Holotype: Divisão de Geologia e Mineralogia, Departamento Nacional de Produção Mineral, Río de Janeiro, Brazil. DGM-DNPM 756-R, part and counterpart of a shell (Price 1973).

Diagnosis: See Meylan (1996) and Gaffney et al. (2006).

Locality, horizon, and age: Gypsum mine, 2 km northeast of Santana do Carri, Ceara, Brazil (Price 1973). The Romualdo Member of the Santana Formation lies between the Crato Member of the Santana Formation and the Exu Formation, which overlies the Santana Formation, Albian, in age (Early Cretaceous) (Maisey 1990, 1991). Fielding et al. (2005) described a new species of *Araripemys* from the Crato Formation (Late Aptian). This species, *Ar. arturi*, was synonymized with *Ar. barretoi* by Gaffney et al. (2006) (Fig. 5.8c, e).

Comments: *Araripemys barretoi* was described by Price (1973: Figs. 1–5) on the basis of a shell. Lyonel Price named the clade Araripemydidae for *Araripemys* and placed it in the suborder Amphychelydia. Gaffney (1972, 1975) noted that Amphychelydia was a paraphyletic taxon that includes cryptodiran and pleurodiran species that lack more derived traits. Following Price (1973) publication and based on the suture of the pelvic girdle to the shell, other authors (Broin 1980, 1988; Schleich 1990; Kischlat and Campos 1990; Hirayama 1991; Meylan and Gaffney 1991; Kischlat 1996; Meylan 1996) assigned *Araripemys* to Pleurodira. Among these researchers, Broin (1980) was the first to assign this genus to Pleurodira. She also suggested that this taxon is closely related to Chelidae.

More recently, France de Broin (1988) proposed referring *Araripemys* to Pelomedusoides, which suggestion was followed by Schleich (1990) when he described an additional shell. More recently, Kischlat and Campos (1990) described axial and appendicular skeletons as well as shell remains. Meylan and Gaffney (1991) have presented photographs and a succinct description of *Araripemys barretoi*. These authors included figures of the first *Araripemys* skull recovered and concluded that this taxon is a pelomedusoid (pelomedusid *sensu lato* in the cited article) on the basis of diagnostic cervical and cranial characters. Meylan (1996) published the first detailed description of *Araripemys barretoi* cranial and postcranial material as well as the first detailed phylogenetic analysis. Meylan concluded that *Araripemys* and *Euraxemys* (as FR 4929 in his article) form a monophyletic clade that is the sister group to all other Pelomedusoides.

Although Broin (1980) referred several shell fragments from the Aptian Gadoufaoua (Niger) to ? *Araripemys*, Gaffney et al. (2006) pointed out that while these materials are similar in texture to *Araripemys*, they do not show derived characters other than the iliac suture on the eighth costal bone and it is therefore unjustified to categorize these materials more accurately than Pleurodira incertae sedis.

Araripemys arturi was described by Fielding et al. (2005) from the Upper Aptian/ Lower Albian Crato Formation south of Nova Olinda in the Araripe Basin, northeastern Brazil. These authors have recognized characters to differentiate *Ar. arturi* from *Ar. barretoi* including shell shape, differences in peripheral bones, and the shape of ungueal phalanges. According to Gaffney et al. (2006), these traits are not enough to differentiate the species, and these authors noted that, in the holotype, two of these supposed differences (e.g., carapace outline, peripheral differences) might be a result of ontogenetic or sexual variation, while the other (shape of terminal phalanges) can be an artifact of preservation. These considerations allow Gaffney et al. (2006) to place *Ar. arturi* as asynonymous of *Ar. barretoi*.[4]

5.2 Discussion

Different groups of organisms that exhibit a trans-Atlantic distribution (Laurent 1979; Goldblatt 1993) have been cited as examples of biological relationships resulting from the past connection between Africa and South America. Accordingly, continental drift and the fragmentation of Gondwana have been cited as important factors in the distribution of several groups of vertebrates (e.g., Sampson et al. 1998; Maisey 2000; Buckley et al. 2000; Turner 2004). Based on abelisaurid's phylogeny, Sampson et al. (1998) suggested a connection between South America, India, and Madagascar during the Upper Cretaceous. A similar pattern based on crocodyliform relationships has been described by Buckley et al. (2000) and Turner (2004). Among turtles, the pleurodiran pelomedusoids have been also considered a primary example

[4]Courtesy of the American Museum of Natural History.

of phylogenetic congruence with the current hypothesis of tectonic evolution (Pritchard 1984; Broin 1988; Bauer 1993).

Two different lineages are included in the pleurodiran clade: Platychelyidae and Eupleurodira. The first is restricted to Oxfordian, Kimmeridgian, Tithonian, and Valanginian species (Bräm 1965; de la Fuente and Fernández 1989; Fernández and de la Fuente 1994; Cadena and Gaffney 2005; Lapparent de Broin et al. 2007), while Eupleurodira [with punctual records of Pelomedusoids in Valanginian (Cadena 2011) and Barremian (Gallo et al. 2009) in South America] has an extended record from Aptian to Recent times (see Gaffney et al. 2006 and references therein). Eupleurodira is divided in two main clades: Chelidae and Pelomedusoides. The first has a disjointed range in Australasia and South America (Iverson 1992; see also Chapter 4), while pelomedusoids have a wider distribution (Wood 1984; Pritchard 1984; Pritchard and Trebbau 1984) with extant species restricted to northern South America, sub-Saharan Africa, and Madagascar (Iverson 1992, Pritchard and Trebbau 1984).

Incongruence between the phylogenetic relationships based on morphological data (Gaffney and Meylan 1988; Gaffney et al. 1991) and molecular data (Seddon et al. 1997; Noonan 2000) is apparent in different analyses of Podocnemididae, which supports alternative biogeographic hypotheses for the clade. This incongruence is also evident in combined molecular and morphological data sets (Shaffer et al. 1997). Morphological data (Gaffney 1988; Gaffney and Meylan 1988) support a close relationship between the two South American genera (*Podocnemis* and *Peltocephalus*), while molecular data have broken up the monophyly of Neotropical taxa and have suggested that *Podocnemis* is the sister group of the Malagasy *Erymnochelys* (Seddon et al. 1997; Noonan 2000; Noonan and Chippindale 2006).

The molecular evidence and reanalysis of Meylan's (1996) data matrix led Brice Noonan to conclude that "current pelomedusoid distribution and relationships could be the result of large scale extinctions with extant taxa being relict of this widespread group" (Noonan 2000: p. 1245). Noonan (2000) explanation was based on Maisey's (1993) statement that "the present distribution of pelomedusids…is clearly a relict of an earlier, more cosmopolitan, post-Gondwanan distribution." As noted by Romano and Azevedo (2006), Noonan's (2000) interpretation failed to consider the parallelism in pelomedusoids described by Wood (1984), Broin (1988), and Lapparent de Broin (2000). According to Romano and Azevedo (2006), this parallelism might explain the incongruence between molecular and morphological data proposed by Noonan (2000).

Romano and Azevedo (2006) reanalyzed de la Fuente (2003) dataset. Based on their phylogenetic analysis and the temporal distribution of taxa, it is possible to conclude that association of vicariant events of pelomedusoid lineages occurred as a consequence of the fragmentation of Gondwana during the Cretaceous. The separation of north Gondwana pelomedusoids and south Gondwana chelids (recognized by Broin 1987, 1988; Broin and de la Fuente 1993) started after the divergence of Chelidae and Pelomedusoides. These two eupleurodiran clades are the result of the first vicariant event recognized by Romano and Azevedo (2006), which resulted in the isolation of Lower Cretaceous pelomedusidae in northern Gondwana (northwestern South America–northeastern Africa) and chelids in Patagonia, Antarctica, and Australia (Broin 1980; Lapparent de Broin 2000). According to Romano and Azevedo (2006), volcanic activity during the Early Cretaceous, represented by basaltic magmatism of the Serra Geral Formation (Renne et al. 1992), might have isolated both lineages of eupleurodires. Differentiation and expansion of the Pelomedusoides had started at least by Aptian times throughout Africa and South America as evident in the fossil record, as it is shown by the earlier records of *Araripemys* and Pelomedusidae date to the Late Aptian in northeastern Brazil (Meylan 1996; Fielding et al. 2005; Oliveira and Kellner 2005) and Late Aptian in Gadoufaoua, Niger (Broin 1980; Lapparent de Broin 2000). Three pelomedusoid lineages from the end of the Lower Cretaceous in northeastern South America are recognized: Euraxemyididae, Bothremydidae, and *Brasilemys*. At this time, the basal members of Podocnemidura (sensu Gaffney et al. 2006) were distributed in northeastern South America and northwestern Africa, while Pelomedusidae (sensu stricto) should have differentiated in Africa.

Africa was isolated from the remaining components of Gondwana supercontinent at the end of the Lower Cretaceous. Consequently, this paleogeographic circumstance and the phylogenetic hypothesis proposed by Romano and Azevedo (2006) prompted these authors to suggest that the next vicariant event occurred at *Hamadachelys*' node, in Romano and Azevedo (2006) phylogeny. The appearance of the other members of Podocnemidura (sensu Gaffney et al. 2006), such as *Portezueloemys*, *Bauruemys*, and Podocnemidididae, supports a southward migration and colonization of southern South America, Madagascar, and Antarctica (Lapparent de Broin 2000; de la Fuente 2003). According to Romano and Azevedo (2006), the connection among Africa, India, and Madagascar via the Kerguelen Plateau (Sampson et al. 1998) during the Middle and Upper Cretaceous probably allowed the passage of podocnemidids from South America to Madagascar through Antarctica. After the isolation of these continental masses, the vicariant event of podocnemidids occurred in South America and Madagascar.

More recently, several authors (Franca and Langer 2006; Meylan et al. 2009; Cadena et al. 2010; Gaffney et al. 2011) have addressed new morphology-based phylogenetic analyses including extinct and extant species. When only the extant species are considered, a different topology is obtained in relation to previous phylogenetic studies based on morphology (i.e., Gaffney 1988; Gaffney and Meylan 1988; Lapparent de Broin 2000; de la Fuente 2003; Romano and Azevedo 2006) (Fig. 5.11: Morpho A). These new morphological studies suggest *Podocnemis* is the sister taxon of *Erymnochelys* plus *Peltocephalus* (Fig. 5.11: Morpho B). Paraphyly of South American podocnemidid with respect to *Erymnochelys* is now supported by morphological data as well as previous molecular data (Georges et al. 1998; Noonan 2000; Noonan and Chippindale 2006). Two recent studies—the analysis of nuclear and mitochondrial DNA sequences of extant podocnemidid species addressed by Vargas-Ramírez et al. (2008) and a total evidence of morphological and molecular analysis yielded by Cadena et al. (2012)—are in agreement with previous molecular studies (Georges et al. 1998; Noonan 2000; Noonan and Chippindale 2006) (Fig. 5.11: Molecules). These phylogenetic results corroborate the

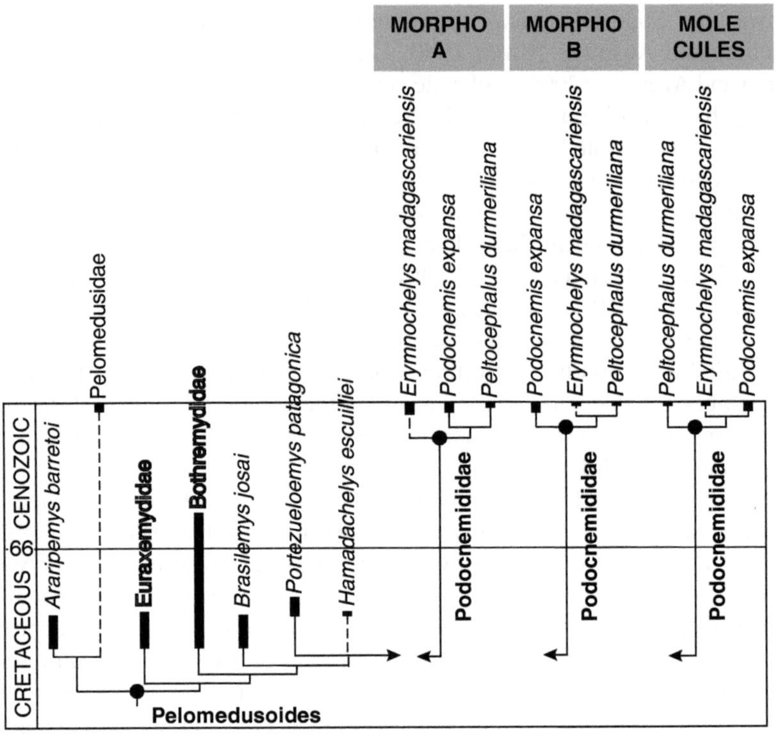

Fig. 5.11 Calibrated trees of competing hypotheses of podocnemidid evolution. The morphological cladogram *A* is based on Gaffney and Meylan (1988) and *B* is based on Gaffney et al. (2011). The molecular cladogram is based on Vargas-Ramírez et al. (2008) and the total evidence analysis of Cadena et al. (2012). Hollow lettering, extinct clade. Fill lettering, crown group. Dashed line, African taxa. Continuous line, South American taxa. Ages in Million Years ago

connection between South America and Madagascar and support the existence of a Late Cretaceous land bridge between Madagascar–India and South America–Antarctica. These findings have been cited as evidence against trans-Atlantic vicariance (Noonan 2000; Noonan and Chippindale 2006). Another fact that challenges the vicariant scenario is the Pangaeic distribution of the extinct clade of bothremydids. This distributional pattern led Maisey (1993) to suggest that the extant podocnemidid species are a relict of a widespread distribution of Pelomedusoides.

References

Alarcon Pardo H (1969) Contribución al conocimiento de la morfología, ecologia, comportamiento y distribución geográfica de *Podocnemis vogli*, Testudinata (Pelomedusidae). Rev Acad Colomb Cien Exac Fis Nat 13:303–329

Andrews CW (1901) Preliminary note on some recently discovered extinct vertebrates from Egypt (Part II). Geol Mag (n ser) 4:436–444

Antunes MT, de Broin F (1988) Le Crétacé terminal de Beira Litoral, Portugal: remarques stratigraphiques et écologiques, étude complémentaire de *Rosasia soutoi* (Chelonii, Bothremydidae). Cienc Terra 9:153–200

Bauer AM (1993) African–South American relationships: a perspective from the Reptilia. In: Goldblatt P (ed) Biological relationships between Africa and South America. Yale University Press, New Haven

Baur G (1888) Osteologische Notizen über Reptilien. (Fortsetzung III). Zool Anzeiger 11:417–424

Baur G (1891) Notes on some little known American fossil tortoises. Proc Acad Nat Sci Phila 43:411–430

Bertini JB, Santucci RM, Toledo CEV, Menegazzo MC (2006) Taphonomy and depositional history of an Upper Cretaceous turtle bearing outcrop from the Adamantina Formation, southwestern São Paulo State. Rev Brasil Paleontol 9:181–186

Bocquentin J, Melo J (2006) *Stupendemys souzai* sp. nov. (Pleurodira, Podocnemididae) from the Miocene-Pliocene of the Solimões Formation, Brazil. Rev Bras Paleontol 9:187–192

Bocquentin J, Negri FR (1993) Sobre la ocorrencia do quelônio gigante Stupendemys(Pleurodira, Podocnemididae, Podocnemidinae) no Mioceno-superior Plioceno da Amazônia sul-ocidental. Ameghiniana 30:324–325

Boulenger GA (1889) Catalogue of the Chelonians, Rhynchocephalians, and Crocodiles in the British Museum (Natural History). Trustees of the British Museum (Natural History), London

Bräm H (1965) Die Schildkröten aus dem oberen Jura (Malm) der Gegend von Solothurn. Schweiz Paläontol Abh 83:1–190

Buckley GA, Brochu CA, Krause DW, Pol D (2000) A pug-nosed crocodyliform from the Late Cretaceous of Madagascar. Nature 405:941–944

Cadena E (2011) Potential Earliest Record of Podocnemidoid Turtles from the Early Cretaceous (Valanginian) of Colombia. J Paleontol 85:877–881

Cadena E, Gaffney ES (2005) *Notoemys zapatocaensis*, a new side–necked turtle (Pleurodira: Platychelyidae) from the Early Cretaceous of Colombia. Am Mus Novit 3470:1–19

Cadena EA, Bloch JI, Jaramillo CA (2010) New podocnemidid turtle (Testudines: Pleurodira) from the Middle–Upper Paleocene of South America. J Vertebr Paleontol 30:367–382

Cadena EA, Ksepka DT, Jaramillo CA, Bloch JI (2012) New pelomedusoid turtles from the Late Palaeocene Cerrejón Formation of Colombia and their implications for phylogeny and body size evolution. J Syst Palaeontol 10:313–331

Candeiro CRA (2009) Vertebrates of the Marília Formation (Late Maastrichtian) from the Peirópolis paleontological site: toward a better understanding. Earth Sci Res J 13:6–15

Candeiro CRA, Martinelli AG, Avilla LS, Rich TH (2006) Tetrapods from the Upper Cretaceous (Turonian-Maastrichtian) Bauru Group of Brazil: a reappraisal. Cretaceous Res 27:923–946

Carrington da Costa J (1940) Um novo quelônio fóssil. Comm Serv Geol Port 21:107–125

Carvalho P, Bocquentin J, Lapparent de Broin F (2002) Une nouvelle espèce de *Podocnemis* (Pleurodira, Podocnemididae) provenant du Neógène de la Formation Solimões, Acre Brésil. Geobios 35:677–686

Cattoi N, Freiberg MA (1958) Una nueva especie de "Podocnemis" del Cretáceo argentino. Physis 21:58–67

Collins RL, Lynn WG (1936) Fossil turtles from Maryland. Proc Am Phil Soc 76:151–174

Cope ED (1868) On the origin of genera. Proc Acad Nat Sci Phila 20:242–300

Cornalia E (1849) Vertebratorum synopsis in Museo Mediolanense extantium quae per no vum Orbem Cajetanas Osculati collegit Annis 1846-47-48. Speciebus novis vel minus cognitus adjectis, nec non Descriptionibus atque Iconibus Illustratis, curante Aemilio Cornalia Mediolani

de Broin F (1971) Une espèce nouvelle de tortue pleurodire (? *Roxochelys vilavilensis* n. sp.) dans le Crétacé supérieur de Bolivie. Bull Soc Géol Fr 7(3–4):445–452

de Broin F (1977) Contribution à l'étude des chéloniens; chéloniens continentaux du Crétace et du Tertiaire de France. Mém Mus Nat Hist Nat Paris Serie C Geologie 38:1–366

de Broin F (1980) Les tortues de Gadoufaoua (Aptien du Niger); aperçu sur la paléobiogéographie des Pelomedusidae (Pleurodira). Mém Soc Géol Fr 139:39–46

de Broin F (1987) The late Cretaceous Fauna of Los Alamitos, Patagonia, Argentina. Part IV Chelonia. Rev Mus Argent Cienc Nat "Bernardino Rivadavia", Paleontol 3:131–139
de Broin F (1988) Les Tortues et le Gondwana. Examen des rapport entre le fractionemment du Gondwana et la dispersión géographique dess tortues pleurodires àpartir de Crétacé. Stud Geol Salmant Stud Palaeochel 2:103–142
de Broin F (1991) Fossil turtles from Bolivia. In: R Suárez-Soruco (ed) Fósiles y facies de Bolivia. Vol. I. Vertebrados, Rev Téc Yac Petrolif Fisc Bolivia, vol 12, pp 509–527
de Broin F, de la Fuente MS (1993) Les tortues fossiles d'Argentine: synthèse. Ann Paléontol 79:169–232
de Campos DA, Kellner AWA (1999) On some sauropod (Titanosauridae) pelves from the continental Cretaceous of Brazil. Nat Sci Mus Monogr Tokyo 15:143–166
de Gasparini ZB, Báez AM (1975) Aportes al conocimiento de la herpetofauna terciaria de la Argentina. Actas 1° Congr Arg Paleontol Bioestr Tucumán 2:377–415
de la Fuente MS (2003) Two new Pleurodiran turtles from the Portezuelo Formation (Upper Cretaceous) of northern Patagonia, Argentina. J Paleontol 77:559–575
de la Fuente MS (2007) Testudines. In: de Gasparini ZB, Coria R, Salgado L (eds) Patagonian Mesozoic reptiles. Indiana University Press, Bloomington
de la Fuente MS, Fernández MS (1989) *Notoemys laticentralis* Cattoi & Freiberg, 1961, from the Upper Jurassic of Argentina: a member of the Infraorder Pleurodira (Cope, 1868). Stud Geol Salmant Stud Palaeochel 3:25–32
de la Fuente MS, Ciccioli P, Limarino O, Gutiérrez P, Fauque L (2003) Quelonios podocnemídidos en la Formación Puesto La Flecha (Oligoceno), Precordillera de la Rioja, Argentina. Ameghiniana 40:617–624
Díaz de Gamero ML, Linares OJ (1989) Estratigrafía y paleontología de la Formación Urumaco del Mioceno Tardío de Falcon noroccidental. 7° Congr Geol Venez Mem 1:419–438
Duméril AMC (1852) Description des reptiles nouveaux ou imparfaitement connus de la collection du Muséum d'Histoire Naturelle, et remarques sur la classification et les charactères des reptiles. Arch Mus Hist Nat Paris 6:209–264
Duméril AMC, Bibron G (1835) Erpétologie génerale du histoire naturelle complète des reptiles. Roret, Paris
Ernst CH, Barbour RW (1989) Turtles of the world. Smithsonian Institution Press, Washington, DC
Fernández MS, de la Fuente MS (1994) Redescription and phylogenetic position of *Notoemys*: The oldest Gondwanian pleurodiran turtle. N Jahrb Geol Paläontol Abh 19:81–105
Fielding S, Martill DM, Naish D (2005) Solnhofen-style soft-tissue preservation in a new species of turtle from the Crato Formation (Early Cretaceous, Aptian) of north-east Brazil. Palaeontology 48:1301–1310
Flynn JJ, Guerrero J, Swisher CC III (1997) Geochronology of the Honda Group. In: Kay RF, Madden RH, Cifelli RL, Flynn JJ (eds) Vertebrate paleontology in the Neotropics. Smithsonian Institution Press, Washington, DC
Frailey CD (1986) Late Miocene and Holocene mammals exclusive of the Notoungulata, of the Río Acre region, western Amazonia. Contrib Sci Nat Hist Mus Los Angel 374:1–46
Franca MAG (2004) Osteologia de um quelônio fôssil (Pleurodira, Podocnemididae) do Cretáceo superior (Maastrichtiano) de Minas Gerais, Brasil. Unpublished Doctoral dissertation, Universidade da Facultade de Filosofia, Ciencias e Letras de Ribeirao Preto de São Paulo, Departamento de Biologia da Facultade de Filosofia, Ciências e Letras de Ribeirão Preto
Franca MAG, Langer MC (2005) A new freshwater turtle (Reptilia, Pleurodira, Podocnemidae) from the Upper Cretaceous (Maastrichtian) of Minas Gerais, Brazil. Geodiversitas 27:391–411
Franca MAG, Langer MC (2006) Phylogenetic relationships of the Bauru Group turtles (Late Cretaceous of south-central Brazil). Rev Brasil Paleontol 9:365–373
Fretey J (1977) Les Cheloniens de Guyane Francaise. 1. Etude preliminaire. Thesis, Université de Paris
Gaffney ES (1972) The systematics of the North American family Baenidae (Reptilia, Cryptodira). Bull Am Mus Nat Hist 147:241–320
Gaffney ES (1975) A phylogeny and classification of the higher categories of turtles. Bull Am Mus Nat Hist 155:389–436

Gaffney ES (1977) The side-necked turtle family Chelidae: a theory of relationships using sharing characters. Am Mus Novit 2620:1–28
Gaffney ES (1979) Comparative cranial morphology of recent and fossil turtles. Bull Am Mus Nat Hist 164:65–376
Gaffney ES (1988) A cladogram of the pleurodiran turtles. Acta Zool Cracov 31:487–492
Gaffney ES (1990) The comparative osteology of the Triassic turtle *Proganochelys*. Bull Am Mus Nat Hist 194:1–263
Gaffney ES, Meylan P (1988) A phylogeny of turtles. In: Benton MJ (ed) The phylogeny and classification of the tetrapods 1: amphibians, reptiles, birds, vol 35A, Systematics association, special volume. Clarendon, Oxford
Gaffney ES, Meylan PA (1991) Primitive pelomedusid turtle. In: Maisey JG (ed) Santana fossils: an illustrated atlas. Tropical Fish Hobbyist Publications, Neptune
Gaffney ES, Tong H (2003) *Phosphatochelys*, a new side-necked turtle (Pelomedusioides: Bothremydidae) from the Paleocene of Morocco. In: Flynn LG (ed) Vertebrate fossils and their context: contributions in honor of Richard H Tedford, vol 279, Bulletin of the American Museum of Natural History. American Museum of Natural History, New York, pp 644–659
Gaffney ES, Wood RC (2002) *Bairdemys*, a new side-necked turtle (Pelomedusoides: Podocnemididae) from the Miocene of the Caribbean. Am Mus Novit 3359:1–28
Gaffney ES, Meylan PA, Wyss A (1991) A computer assisted analysis of the relationships of the higher categories of turtles. Cladistics 7:313–335
Gaffney ES, Campbell KE, Wood RC (1998) Pelomedusoid side-necked turtles from Late Miocene Sediments in southwestern Amazonia. Am Mus Novit 3245:1–12
Gaffney ES, Chatterjee S, Rudra DK (2001a) *Kurmademys*, a new side-necked turtle (Pelomedusoides: Bothremydidae) from the Late Cretaceous of India. Am Mus Novit 3321:1–16
Gaffney ES, de Almeida Campos D, Hirayama R (2001b) *Cearachelys*, a new side-necked turtle (Pelomedusoides: Bothremydidae) from the Early Cretaceous of Brazil. Am Mus Novit 3319:1–20
Gaffney ES, Moody RTJ, Walker CA (2001c) *Azabbaremys*, a new side-necked turtle (Pelomedusoides: Bothremydidae) from the Paleocene of Mali. Am Mus Novit 3320:1–16
Gaffney ES, Tong H, Meylan PA (2002) *Galianemys*, a new side-necked turtle (Pelomedusoides: Bothremydidae) from the Late Cretaceous of Morocco. Am Mus Novit 3379:1–20
Gaffney ES, Sahni A, Schleich H, Singh SD, Srivastava R (2003) *Sankuchemys*, a new side-necked turtle (Pelomedusoides: Bothremydidae) from the Late Cretaceous of India. Am Mus Novit 3405:1–10
Gaffney ES, Tong H, Meylan PA (2006) Evolution of the side-necked turtles: the families Bothremydidade, Euraxemydidae, and Araripemydidae. Bull Am Mus Nat Hist 300:1–698
Gaffney ES, Scheyer TM, Johnson KG, Bocquentin Villanueva J, Aguilera OA (2008) Two new species of the side necked turtle genus, Bairdemys (Pleurodira, Podocnemididae), from the Miocene of Venezuela. Palaeontol Zeit 82:209–229
Gaffney ES, Meylan PA, Wood RC, Simons E, De Almeida Campos D (2011) Evolution of the side-necked turtles The Family Podocnemididae. Bull Am Mus Nat Hist 350:1–237
Gallo VG, Silva HMA, Ramo RRC, Petra R, Romano PSR (2009) A tartaruga Pleurodira mais antigua do Brasil. An 21 Congr Brasil Paleontol, Belém.18- PA 180–181
Gayet M, Marshall LG, Sempere T (1991) The Mesozoic and Paleocene Vertebrates of Bolivia and their Stratigraphic context: a review. In: Suárez-Soruco R (ed) Fósiles y Facies de Bolivia, vol 12, Revista técnica de YPFB. Yacimientos Petrolíferos Fiscales Bolivianos, Santa Cruz, pp 393–434
Georges A, Birrell J, Saint KM, McCord WP, Donnellan SC (1998) A phylogeny for side-necked turtles (Chelonia: Pleurodira) based on mitochondrial and nuclear gene sequence variation. Biol J Linn Soc 67:213–246
Goldblatt P (1993) Biological relationships between Africa and South America. Yale University Press, New Haven
Gómez Omil RJ, Boll A, Hernández RM (1989) Cuenca Cretácico-Terciario Del Noroeste Argentino (Grupo Salta). In: Chebli G, Spalletti LA (eds) Cuencas sedimentarias Argentinas, vol 6, Serie Correl Geol., pp 43–64
Groombridge B (1982) The IUCN Amphibia-Reptilia Red Data Book. Testudines. Crocodylia. Rhynchocephalia Part 1. IUCN, Gland

Guerrero J (1997) Stratigraphy, sedimentary environments, and the Miocene uplift of the Colombian Andes. In: Kay RF, Madden RH, Cifelli RL, Flynn JJ (eds) Vertebrate paleontology in the Ne tropics. Smithsonian Institution Press, Washington, DC

Hirayama R (1991) Phylogenetic relationship of *Araripemys* (family Araripemydidae; Pleuro Pleurodira; Testudinata). Abstracts of the 140th regular meeting of the Palaeontological Society of Japan (June 22, 23, 1991, Chiba):34

Hoogmoed MS, Grüber U (1983) Spix and Wagler type specimens of reptiles and amphibians in the Natural History Museum in Munich (Germany) and Leiden (The Netherlands). Spixiana Suppl 9:319–415

Iverson JB (1992) A revised checklist with distribution maps of the turtles of the world. Privately Printed, Richmond

Jaramillo CA, Bayona G, Pardo A, Rueda MJ, Harrington G, Mora G (2007) Palynology of the Upper Palaeocene Cerrejón Formation, Northern Colombia. Palynology 31:153–189

Kay RF, Madden RH, Cifelli RL, Flynn JJ (1997) Vertebrate paleontology in the Neotropics. Smithsonian Institution Press, Washington, DC

Kellner AWA, de Almeida Campos D, Trotta MNF (2005) Description of a titanosaurid caudal series from the Bauru Group, Late Cretaceous of Brazil. Arq Nac Rio de Janeiro 63:529–564

King FW, Burke RL (1989) Crocodilian, Tuatara and Turtle species on the World. Association of Systematics Collections, Washington, DC

Kischlat EE (1994) Observações sobre *Podocnemis elegans* Suarez (Chelonii, Pleurodira, Podocnemididae) do Neocretáceo do Brasil. Acta Geol Leopoldensia 17:345–351

Kischlat EE (1996) Preliminary phylogenetic analysis of the pleurodiran chelonians from the Cretaceous of Brazil. An Acad Brasil Ciên 68:288–289

Kischlat EE, de Almeida Campos D (1990) Some osteological aspects of *Araripemys barretoi* Price, 1973 (Chelonii, Pleurodira, Araripemydi dae). In: 1° Simposio Sobre A Bacia Do Araripe E Bacias Interiores Do Nordeste, 1990. Actas: 387–395

Lapparent de Broin F (2000) The oldest pre-Podocnemidid turtle (Chelonii, Pleurodira), from the Early Cretaceous, Ceará state, Brazil, and its environment. Treb Mus Geol Barcelona 9:43–95

Lapparent de Broin F, Werner C (1998) New Late Cretaceous turtles from the Western Desert, Egypt. Ann Paléontol 84:131–214

Lapparent de Broin F, Bocquentin J, Negri FR (1993) Gigantic turtles (Pleurodira, Podocnemididae) from the Late Miocene–Early Pliocene of southwestern Amazon. Bull Inst Francais Études Andines 22:657–670

Lapparent de Broin F, de la Fuente MS, Fernández MS (2007) *Notoemys* (Chelonii, Pleurodira), Late Jurassic of Argentina: new examination of the anatomical structures and comparisons. Rev Paléobiol 26:99–136

Laurent RF (1979) Herpetofaunal relationships between Africa and South America. In: Duellman WE (ed) The South American herpetofauna: its origin, evolution and dispersal, vol 7, Monograph of the Museum of Natural History, the University of Kansas. Museum of Natural History, University of Kansas, Lawrence, pp 55–71

Leanza HA (1999) The Jurassic and Cretaceous terrestrial beds from Southern Neuquén Basin, Argentina. Field Guide Inst Sup Correl Geol Miscelánea 4:1–30

Leanza HA, Apesteguía S, Novas FE, de la Fuente MS (2004) Cretaceous terrestrial beds from Neuquén Basin (Argentina) and their tetrapods assemblages. Cretaceous Res 25:61–87

Linares OJ (2004) Bioestratigrafía de la fauna de mamíferos de las Formaciones Socorro, Urumaco y Codore (Mioceno Medio-Plioceno Temprano) de la Formación de Urumaco, Falcón, Venezuela. Paleobiol Neotr 2:1–14

Luederwaldt H (1926) Os Chelonios brasileiros com a lista das especies do Museu Paulista. Rev Mus Paulista 14:404–468

Maisey JG (1990) Stratigraphy and depositional environment of the Crato Member (Santana Formation, Lower Cretaceous) of N.E. Brazil. In: Grimaldi DA (ed) Insects from the Santana Formation, Lower Cretaceous, of Brazil, vol 195, Bulletin of the American Museum of Natural History. American Museum of Natural History, New York, pp 15–19

Maisey JG (1991) Santana fossils: an illustrated atlas. Tropical Fish Hobbyist Publications, Neptune

Maisey JG (1993) Tectonics, the Santana Lagerstatten, and the implications for late Gondwanan biogeography. In: Goldblat P (ed) Biological relationships between Africa and South America. Yale University Press, New Haven

Maisey JG (2000) Continental break up and the distribution of fishes of Western Gondwana during the Early Cretaceous. Cretaceous Res 21:281–314

Marshall LG, Hoffstetter R, Pascual R (1983) Mammals and Stratigraphy: geochronology of the continental mammal-bearing tertiary of South America. Palaeovertebrata Mém Extr 1:1–93

Matiazzi W (2007) Redescrição da anatomia craneana e pós-craneana do fóssil *Bauruemys elegans* (Suárez 1969) do Cretáceo Superior de Bacia da Bauru e seu posicionamento filogenético dentro da epifamilia Podocnemidinura (Testudines Pelomedusoides) Master Dissertation Universidade de São Paulo

Meylan PA (1996) Skeletal morphology and relationships of the Early Cretaceous side necked turtle, *Araripemys barretoi* (Testudines: Pelomedusoides: Araripemydidae), from the Santana Formation of Brazil. J Vertebr Paleontol 16:20–33

Meylan PA, Gaffney ES (1991) *Araripemys* Price, 1973. In: Maisey JG (ed) Santana fossils: an illustrated atlas: 326–334. Tropical Fish Hobbyist Publications, Neptune

Meylan PA, Gaffney ES, de Almeida Campos D (2009) *Caninemys*, a new side-necked turtle (Pelomedusoides: Podocnemididae) from the Miocene of Brazil. Am Mus Novit 3639:1–26

Mittermeier RA, Wilson RA (1974) Redescription of *Podocnemis erythrocephala* (Spix, 1824), an Amazonian pelomedusid turtle. Pap Avul Zoologia 28:147–162

Moreno JA (1970) Estratigrafía y Paleogeografía del Cretácico Superior en la Cuenca del Noroeste Argentino, con especial mención de los Subgrupos Balbuena y Santa Bárbara. Rev Asoc Geol Arg 25:9–44

Müller L (1935) Über eine neue Podocnemis-Art (*Podocnemis vogli*) aus Venezuela nebst ergänzenden Bemerkungen über die systematischen Merkmale der ihr nächstverwandten Arten. Zool Anzeiger 110:97–109

Neill WT (1965) Notes on the five Amazonian species of *Podocnemis* (Testudinata: Pelomedusidae). Herpetologica 21:287–294

Noonan BP (2000) Does the phylogeny of pelomedudoid turtles reflect vicariance due to continental drift? J Biogeogr 27:1245–1249

Noonan BP, Chippindale PT (2006) Vicariant origin of Malagasy reptiles supports Late Cretaceous Antarctic land bridge. Am Nat 168:730–741

Nopcsa F (1931) Sur des nouveaux restes de tortues du Danien du Midi de la France. Bull Soc Geól Fr 1:223–236

Oliveira GR, Kellner AW (2005) Note on a plastron (Testudines, Pleurodira) from the Lower Cretaceous Crato Member, Santana Formation. Arq Mus Nac 63:523–528

Oliveira GR, Romano PSR (2007) Histórico dos achados de tartarugas fósseis do Brasil. Arq Mus Nac Rio de Janeiro 65:113–133

Pascual R, Bond M, Vucetich MG (1981) El Subgrupo Santa Bárbara (Grupo Salta) y sus vertebrados. Cronología, paleoambientes y paleobiogeografía. Actas 3ª Congr Geol Argent San Luis 3:743–758

Pascual R, Díaz de Gamero ML (1969) Sobre la presencia del género *Eumegamys* (Rodentia Caviomorpha) en la Formación Urumaco del Estado Falcón (Venezuela). Su significación cronológica. AsocVen Geol Min Petr Bol Inf 12:377–388

Pascual R, Vucetich MG, Fernández J (1978) Los primeros mamíferos (Notoungulata, Henricosborniidae) de la Formación Mealla (Grupo Salta, Subgrupo Santa Barbara). Sus implicancias filogenéticas, taxonómicas y cronológicas. Ameghiniana 15:366–390

Portis A (1882) Les Chéloniens de la Molasse vaudoise conservés dans le Musée Géologique de Lausanne. Appendice, Sur un nouveau genre de Chéloniens fossiles de la Craie supérieure de Fuveau. Mém Soc Paléontol Suisse 9:65–76

Price LI (1953) Os quelônios da Formaçâo Bauru, Cretáceo terrestre do Brasil meridional. Bol Div Geol Miner Depart Nac Prod Min 147:1–34

Price LI (1973) Quelônio amphichelydia no Cretáceo inferior do nordeste do Brazil. Rev Brasil Geociências 3:84–96

Pritchard PCH (1967) Living turtles of the world. Tropical Fish Hobbyist Publications, Jersey City

Pritchard PCH (1979) Encyclopedia of turtles. Tropical Fish Hobbyist Publications, Neptune

Pritchard PCH (1984) Evolution and zoogeography of South American turtles. Stud Geol Salmant Vol Esp 1 Stud Palaeochel 1:225–233

Pritchard PCH, Trebbau P (1984) The turtles of Venezuela. SSAR Contrib Herpetology, Ithaca

Quiroz LI, Jaramillo CA (2010) Stratigraphy and sedimentary environments of Miocene shallow to marginal marine deposits in the Urumaco trough, Falcon Basin, Western Venezuela. In: Sánchez-Villagra MR, Aguilera OA, Carlini AA (eds) Urumaco and Venezuelan paleontology. Indiana University Press, Bloomington

Ramo C (1982) Biología de Galápago (*Podocnemis vogli* Muller 1935) en el Hato El Frio llanos de Apure, Venezuela. Doñana Acta Vert 9:1–161

Renne PR, Ernesto M, Pacca LG, Coe RS, Glen JM, Prevot M, Richards MA (1992) The age of the Parana flood volcanism, rifting of Gondwanaland and the Jurassic Cretaceous boundary. Science 258:975–979

Romano PS, Azevedo SAK (2006) Are extant podocnemidid turtles relicts of a wide spread Cretaceous ancestor? South Am J Herpetol 1:175–184

Romano PS, Azevedo SAK (2007) Morphometric analysis of the Upper Cretaceous Brazilian side-necked turtle *Bauruemys elegans* (Suárez, 1969) (Pleurodira, Podocnemididae). Arq Mus Nac Rio de Janeiro 65:395–402

Romano PS, Oliveira GR, Azevedo SAK, Campos DA (2009) Lumping the podocnemidid turtles species from Bauru Basin (Upper Cretaceous of southeastern of Brazil). In: Gaffney Turtle Symposium (October 17–18, 2009, Royal Tyrrell Museum, Drumheller, Canada), Extended Abstracts Vol:141–152

Romano PS, Oliveira GR, Azevedo SAK, Kellner AWA, de Almeida Campos D (2013) New Information about Pelomedusoides(Testudines: Pleurodira) from the Cretaceous of Brazil In: Brinkman DB, Holroyd PA, Gardner JD (eds) Morphology and evolution of turtles, vol 16, pp 261–275. Proceedings of Gaffney Turtle symposium (2009) in Honor of Eugene Gaffney, Vertebrate Paleobiology and Paleoantropology. Springer, Dordrecht

Sampson SD, Witmer LM, Forster CA, Krause DW, O'Connor PM, Dodson P, Ravoavy F (1998) Predatory dinosaur remains from Madagascar: implications for the Cretaceous biogeography of Gondwana. Science 280:1048–1051

Sánchez-Villagra MR, Scheyer TM (2010) Fossil turtles from the northern neotropics: the Urumaco sequence fauna and finds from other localities in Venezuela and Colombia. In: Sánchez-Villagra MR, Aguilera OA, Carlini AA (eds) Urumaco and Venezuelan paleontology. Indiana University Press, Bloomington

Sánchez-Villagra MR, Winkler JD (2006) Cranial variation in *Bairdemys* turtles (Podocnemididae: Miocene of the Caribbean region) and description of new material from Urumaco, Venezuela. J Syst Palaeontol 4:241–253

Scheyer TM, Sánchez-Villagra MR (2007) Carapace bone histology in the giant pleurodiran turtle *Stupendemys geographicus*: phylogeny and function. Acta Palaeontologica Polonica 52:137–154

Schleich HH (1990) Neues Material zu *Araripemys barretoi* Price 1973 (Testudines: Pleurodira). Mitteilungen der Bayer. Staatssamm. Paläontol Hist Geol 30:39–49

Schweigger AF (1812) Prodomus Monographiae Chelooniorum. Königsberg Arch Naturwiss Math 1:171–368

Seddon JM, Georges A, Baverstock PR, McCord W (1997) Phylogenetic relationship of chelid turtles (Pleurodira: Chelidae) based on mitochondrial 12S rRNA sequence variation. Mol Phylogenet Evol 7:55–61

Sereno PC, ElShafie SJ (2013) A new long-necked turtle, *Laganemys tenerensis* (Pleurodira Araripemydidae), from the Elrhaz Formation (Aptian-Albian) of Niger. In: Brinkman DB, Holroyd PA, Gardner JD (eds) Morphology and evolution of turtles, vol 14, pp 215–250. Proceedings of Gaffney Turtle Symposium (2009) in Honor of Eugene Gaffney, Vertebrate Paleobiology and Paleoantropology. Springer, Dordrecht

Shaffer HB, Meylan P, Mcknight ML (1997) Test of turtle phylogeny: molecular, morphological, and paleontological approaches. Syst Biol 46:235–268

Siebenrock F (1902) Zur Systematik der Schildkröten-Gattung *Podocnemis* Wagler. Sitzung richte Kaiserlichen Akad Wissen Wien Mathematisch Naturwissen Klasse 111:157–170

Siebenrock F (1904) Schildkröten von Brasilien. Denks Kaiser Akad Wissen Math Natur Wisse Math Natur Sens Klasse Wien 76:1–28
Staesche K (1937) Podocnemis brasiliensis n. sp. aus der Obern Kreide Brasiliens. N Jahrb Min Geol Paläeontol 77:291–309
Suárez JM (1969) Um quelônio da Formação Baurú. Bol Fac Filosofía Ciên Letras Presidente Prudente 2:35–54
Suárez JM (2002) Sítio fossilífero de Pirapozinho São Paulo: extraordinário depósito de quelônios do Cretáceo. In: Schobbenhaus C, Campos DA, Queiroz ET, Winge M, Berberet-Born MLC (eds) Sítios geológicos e paleontológicos do Brasil. Departamento Nacional da Producçã Mineral, Brasilia, pp 49–54
Swinton WE (1939) A new fresh-water tortoise from Burma. Rec Geol Surv India 74:548–551
Tong H (1998) Pleurodiran turtles from the Eocene of Saint Papoul (Aude), southern France. Oryctos 1:43–53
Tong H, Gaffney ES, Buffetaut E (1998) *Foxemys*, a new side-necked turtle (Bothremydidae: Pelomedusoides) from the Late Cretaceous of France. Am Mus Novit 3251:1–19
Troschel FH (1848) Amphibien. In: Schom burgk MR (ed) Reisen in Britisch-Guiana in den Jahren 1840–44. Im Auftrage Majestät des Königs von Preussen ausgefü hrt. Versuch einer Zusammenstellung der Fauna und Flora von Britisch-Guiana, Leipzig
Turner AH (2004) Crocodyliforms biogeography during the Cretaceous: evidence of Gondwanan vicariance from biogeographical analysis. Proc R Soc Lond B 271:2003–2009
Vanzolini PE (1977) A brief biometrical note on the reproductive biology of some South American Podocnemis (Testudines: Pelomedusiidae). Pap Avul Zool 31:79–102
Vargas-Ramírez M, Castaño-Mora OV, Fritz U (2008) Molecular phylogeny and divergence times of ancient South American and Malagasy river turtles (Testudines: Pleurodira: Podocnemididae). Org Diver Evol 8:388–398
von Spix JB (1824) Animalia nova species novae Testudinum et Ranarum quas in itinere per Brasiliam ann is 1817–1820 collegit et descripsit. Hubschmann, München
Wagler J (1830) Natürliches System der Amphibien, mit Vorangehander Classification der Säugthiere und Vogel. J G Cotta'schen Buchhandlung, München
Weems RE, Knight JL (2013) A new species of *Bairdemys* (Pelomedusoides: Podocnemididae) from the Oligocene (Early Chattian) Chandler Bridge Formation of South Carolina. U.S.A., and its Paleobiogeographic implications for the Genus. In: Brinkman DB, Holroyd PA, Gardner JD (eds) Morphology and evolution of turtles, vol 18, pp 289–303. Proceedings of Gaffney Turtle symposium (2009) in Honor of Eugene Gaffney, Vertebrate Paleobiology and Paleoanthropology. Springer, Dordrecht
Wermuth H, Mertens R (1961) Schildkroten, Krokodile, Brückenechsen. G. Fischer, Jena
Wermuth H, Mertens R (1977) Liste der rezenten Amphibien und Reptilien. Testudines, Crocodylia, Rhynchocephalia. Das Tierreich 100:1–174
Williams EE (1954a) New or redescribed pelomedusid skulls from the Tertiary of Africa and Asia (Testudines, Pelomedusidae) 1. *Dacquemys paleomorpha*, new genus, new species from the Lower Oligocene of the Fayum, Egypt. Breviora 35:1–9
Williams EE (1954b) A key and description of the living species of the genus *Podocnemis* (Sensu Boulenger) (Testudines, Pelomedusidae). Bull Mus Comp Zool 3:279–295
Williams EE (1956) *Podocnemis bassleri*, a new species of pelomedusid turtle from the Late Tertiary of Peru. Am Mus Novit 1782:1–10
Wood RC (1976) *Stupendemys geographicus*, the world's largest turtle. Breviora 436:1–31
Wood RC (1984) Evolution of the Pelomedusid turtles. Stud Geol Salmant Vol Esp 1 Stud Palaeochel 1:269–282
Wood RC (1997) Turtles. In: Kay RF, Madden RH, Cifelli RL, Flynn JJ (eds) Vertebrate paleontology in the Ne tropics. Smithsonian Institution Press, Washington, DC
Wood RC (2003) Fossil turtles from Lothagam. In: Leakey MG, Harris JM (eds) Lothagam: the dawn of humanity in Eastern Africa. Columbia University Press, New York, pp 115–136
Wood RC, de Gamero D (1971) Podocnemis venezuelensis, a new fossil pelomedusid (Testudines, Pleurodira) from the Pliocene of Venezuela and a review of the history of Podocnemis in South America. Breviora 376:1–23

Chapter 6
Meiolaniforms: An Extinct Lineage of Turtles of Gondwanan Origin

Abstract Meiolaniformes include the bizarre group of extinct Cenozoic turtles bearing cranial horns and frills (clade Meiolaniidae) and several species located in its stem. Most meiolaniforms are from Gondwana (Australia and South America), with the exception of two Eurasiatic taxa (*Kallokibotion bajazidi* and *Mongolochelys efremovi*). Meiolaniformes were originated in the present-day South America and they are known from the Lower Cretaceous until the Holocene. The oldest fossils within the clade (*Chubutemys copelloi* and *Otwayemys cunicularius*) indicate a minimum age of origin in the Aptian–Albian (Lower Cretaceous).

Keywords Meiolaniformes • Gondwana • Cretaceous • Paleogene • Phylogeny • Biogeography

6.1 Meiolaniformes Sterli and de la Fuente 2012

The stem-based definition of Meiolaniformes was recently created by Sterli and de la Fuente (2012) to include all the taxa more related to *Meiolania platyceps* than to Cryptodira and Pleurodira. Meiolaniformes includes Meiolaniidae (the bizarre group of extinct Cenozoic turtles bearing cranial horns and frills, caudal rings, and tail club) and several species located in its stem. At minimum, the meiolaniform clade spans the period from the Lower Cretaceous (*Chubutemys copelloi*, Gaffney et al. 2007) to the Holocene (?*Meiolania damelipi*, White et al. 2010) in Asia, Europe, Australia, and South America. Two episodes of clade diversification are shown in the fossil record, the most recent during the Pleistocene–Holocene and the oldest during the Upper Cretaceous.

Genera included: *Chubutemys* Gaffney, Rich, Vickers-Rich, Constantine, Vacca and Kool 2007, *Kallokibotion* Nopcsa 1923, *Meiolania* Owen 1886, *Mongolochelys* Khosatzky 1997, *Ninjemys* Gaffney 1992, *Niolamia* Ameghino 1899, *Otwayemys*

M.S. de la Fuente et al., *Origin, Evolution and Biogeographic History of South American Turtles*, Springer Earth System Sciences, DOI 10.1007/978-3-319-00518-8_6,

Gaffney, Kool, Brinkman, Rich, Vickers-Rich 1998, *Patagoniaemys* Sterli and de la Fuente 2011a, *Peligrochelys* Sterli and de la Fuente 2012, and *Warkalania* Gaffney, Archer and White 1992.

6.1.1 Meiolaniidae Boulenger 1887

Meiolaniids are a peculiar group of extinct turtles bearing cranial horns and frills, caudal rings, and tail clubs and restricted to South America (Eocene) and Australasia (Oligocene to Holocene). Since the discovery of the first meiolaniid, *Meiolania platyceps* (Owen 1886), the relationships between meiolaniids and other turtles have been controversial. Meiolaniids have been considered related to both living groups of turtles, Pleurodira (Boulenger 1887, 1889) and Cryptodira (Huxley 1887; Baur 1889a, 1889b; Gaffney et al. 1991; Gaffney 1996), and even to be outside the crown group (Joyce 2007; Sterli 2010; Sterli and de la Fuente 2011a, b).

The greatest diversity of Meiolaniidae is found in Australasia where the oldest remains are represented by *Warkalania carinaminor* (?Late Oligocene–Early Miocene of Riverside Station, northwest Queensland) and *Meiolania brevicollis* Megirian 1992 (Middle Miocene of Camfield Beds, Northern Territory) (Gaffney 1996). There is a gap in the Australasian record during the Pliocene, but four species of Meiolaniids are present in the Pleistocene–Holocene fossil record of the main continent and surrounding islands: *Ninjemys oweni* (Pleistocene of southern Queensland), *Meiolania platyceps* Owen 1886 (Lord Howe Island), *Meiolania mackayi* Anderson 1925 (Walpole Island, New Caledonia), and ?*Meiolania damelipi* White et al. 2010 (from Efate Island, Vanuatu) (Owen 1886; Anderson 1925; White et al. 2010). Conversely, there is little diversity among named meiolaniids in South America. According to Gaffney (1996) and Sterli and de la Fuente (2011b), there is only one species, the Patagonian *Niolamia argentina*, which is probably Eocene in age.

Genera included: *Niolamia* Ameghino 1899, *Ninjemys* Gaffney 1992, *Meiolania* Owen 1886, and *Warkalania* Gaffney, Archer and White 1992.

South American Taxa

6.1.1.1 *Niolamia* Ameghino 1889

Type species: *Niolamia argentina* Ameghino 1889.
Emended Diagnosis: See Sterli and de la Fuente (2011b).
Species included: Only *Niolamia argentina* Ameghino 1899.

Niolamia argentina Ameghino 1899 (Fig. 6.1)

Neotype: Museo de La Plata, Argentina. MLP 26–40, a skull (Simpson 1938).
According to Sterli and de la Fuente (2011b), a lower jaw housed in the same

Fig. 6.1 Life reconstruction of *Niolamia argentina* from the Paleogene of Argentina

collection (no number) to be part of the neotype the occlusion with upper jaw is perfect.

Referred specimens: MLP 26–41, left scapula; 21–42, right scapula; 21–43, carapace remains; 21–44, tail ring.

Emended Diagnosis: See Sterli and de la Fuente (2011b).

Locality and age: Cañadón Blanco (following MLP collection label information) and Cañadón Hondo localities, Chubut province, Argentina. Upper Cretaceous (following Woodward 1901) or Paleogene (according to the age assigned to the mammalian fauna found in Cañadón Blanco locality (Simpson 1937) (Fig. 6.2a, c).

Comments: As described by Sterli and de la Fuente (2011b), *Niolamia argentina* specimens from the La Plata Museum have an enigmatic history; little is known about the exact location or age of the sediments in which they were found. The exact location and age of its junior synonymous species, *Crossochelys corniger*, are known, however. *Crossochelys corniger* was recovered by G. G. Simpson in the Cañadón Hondo locality (Patagonia, Argentina) in the Eocene outcrops of the Sarmiento Formation. Simpson (1937, 1938) also named and described this species in detail. In several papers, Gaffney (1983, 1996) suggested *Cro. corniger* was a juvenile and declared it as a synonym of *Ni. argentina*, a statement corroborated by Sterli and de la Fuente (2011b).

Although the phylogenetic position of the clade Meiolaniidae is controversial (Gaffney 1996; Hirayama et al. 2000; Gaffney et al. 2007; Joyce 2007), relationships within the clade are well established. In each of the two main competing phylogenetic hypotheses of turtle evolution, *Niolamia argentina* is described as the sister clade of all remaining Australasian meiolaniids (Gaffney 1996; Gaffney et al. 2007; Sterli and de la Fuente 2012). The basal-most position of *Ni. argentina* among meiolaniids is consistent with the stratigraphic record of the group, as this taxon is the oldest member of the clade.

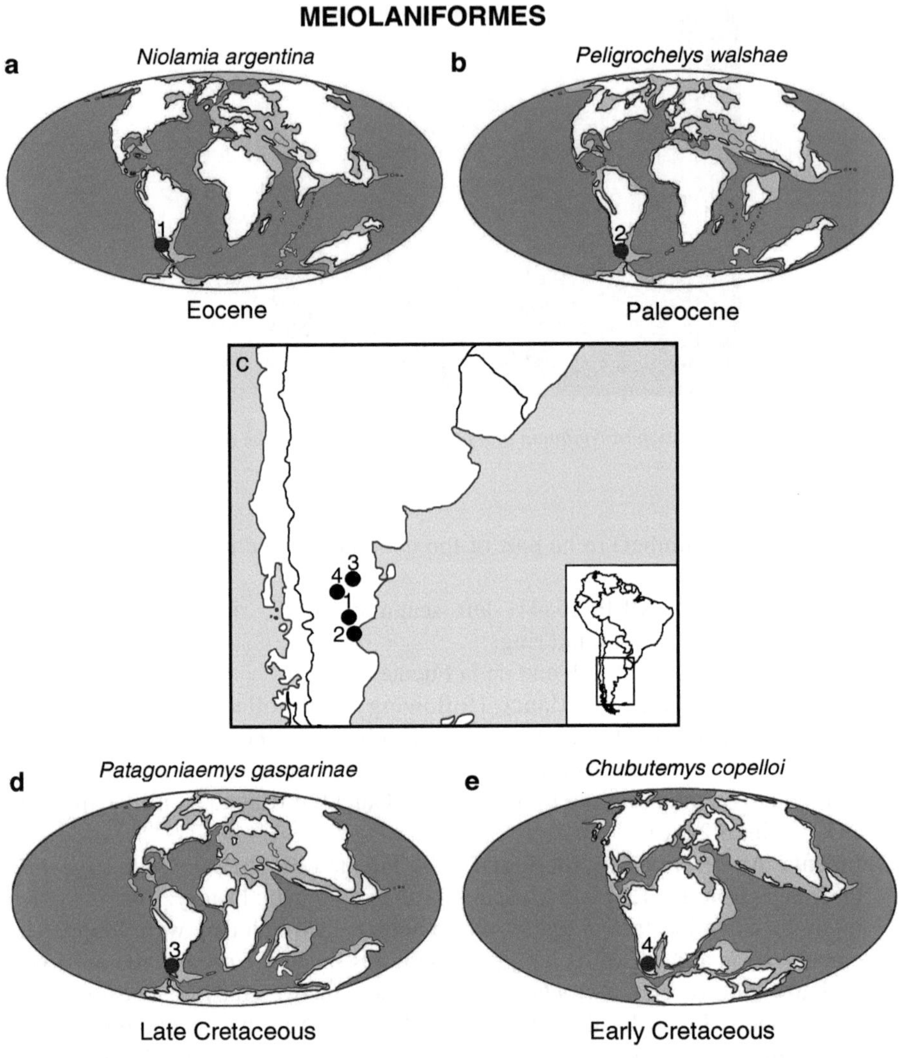

Fig. 6.2 Geographic and stratigraphic distribution of Meiolaniformes in South America. (**a**) geographic distribution of *Niolamia argentina* (*1*) in a paleogeographic reconstruction of continents during the Eocene. (**b**) geographic distribution of *Peligrochelys walshae* (*2*) in a paleogeographic reconstruction of continents during the Palaeocene. (**c**) geographic and stratigraphic location of Meiolaniformes in South America. (**d**) geographic distribution of *Patagoniaemys gasparinae* (*3*) in a paleogeographic reconstruction of continents during the Late Cretaceous. (**e**) geographic distribution of *Chubutemys copelloi* (*4*) in a paleogeographic reconstruction of continents during the Early Cretaceous

6.1.2 Meiolaniforms No Meiolaniidae

South American Taxa

6.1.2.1 *Peligrochelys* Sterli and de la Fuente 2012

Type species: *Peligrochelys walshae* Sterli and de la Fuente 2012.
Diagnosis: See Sterli and de la Fuente (2012).
Species included: Only *Peligrochelys walshae* Sterli and de la Fuente 2012.

Peligrochelys walshae Sterli and de la Fuente 2012

Holotype: MACN PV CH 2017, several skull bones (basisphenoid, basioccipital, right quadrate, right prootic, left opisthotic, a fragment of the right opisthotic, left exoccipital, right frontal, both postorbitals, fragment of left maxilla, fragments of both squamosals, and other skull roof) and a left prezygapophysis.
Diagnosis: See Sterli and de la Fuente (2012).
Locality and age: Punta Peligro locality, 45 km north of the city of Comodoro Rivadavia, Chubut Province, Argentina. Banco Negro Inferior, Salamanca Formation, Danian in age (Lower Palaeocene) (Fig. 6.2b, c).
Comments: *Peligrochelys walshae* was recently named by Sterli and de la Fuente (2012) on the basis of skull remains (see above). In a recent publication, the authors attempted to assess the primary homology between the skull scale pattern of *Peligrochelys walshae*, other meiolaniform taxa, and other basal Testudinata relative to the well-known pattern recognized for Meiolaniidae (Simpson 1938; Gaffney 1996). However, Sterli and de la Fuente (2012) observed that all members of the sister clade of Meiolaniidae have poorly preserved skull remains (e.g., *Patagoniaemys gasparinae* and *Otwayemys cunicularius*) or the skull scale patterns were not described or illustrated in detail (e.g., *Kallokibotion bajazidi*). Nonetheless, the preservation of scales in *Mongolochelys efremovi* and *Peligrochelys walshae* allowed the authors to trace the scale pattern described for meiolaniids in these taxa. Meiolaniidae, *Mongolochelys efremovi*, and *Pel. walshae* share the presence of single scales Y, X, and A along the midline and paired scales C, D, E, G, H, and K. Contrarily to Meiolaniidae, in *Mo. efremovi* and *Pel. walshae* scales F, I, J, and Z are composed of several small scales.

Sterli and de la Fuente's (2012) cladistic analysis resulted in the recovery of Meiolaniidae deeply nested in a monophyletic group dominated by Gondwana taxa (e.g., *Chubutemys copelloi*, *Patagoniaemys gasparinae*, *Otwayemys cunicularius*, *Pel. walshae*), but including some Laurasia representatives (*Mongolochelys efremovi* and *Kallokibotion bajazidi*) as well.

6.1.2.2 *Patagoniaemys* Sterli and de la Fuente 2011a

Type species: *Patagoniaemys gasparinae* Sterli and de la Fuente 2011a.
Diagnosis: See Sterli and de la Fuente (2011a).
Species included: Only *Patagoniaemys gasparinae* Sterli and de la Fuente 2011a.

Patagoniaemys gasparinae Sterli and de la Fuente 2011a

Holotype: Museo Paleontológico Egidio Feruglio, Trelew, Argentina. MPEF-PV 3283. Postorbital bone, a partial carapace, right hyoplastron, right hypoplastron, four cervical vertebrae (2–4, 7), all thoracic vertebrae, 2 sacral vertebrae, 16 caudal vertebrae, pelvic girdles, and distal end of right fibula.
Diagnosis: See Sterli and de la Fuente (2011a).
Locality and age: Buitre Chico Hill, between Bajada Moreno and Bajada del Diablo villages, central Chubut Province, Argentina. La Colonia Formation (Pesce 1979), Campanian–Maastrichtian (Pascual et al. 2000) (Fig. 6.2c, d).
Comments: *Patagoniaemys gasparinae* holotype preserves two opisthocoelus (2nd and 3rd) and two cervical vertebrae, one biconvex (4th) and one biconcave (7th). Sterli and de la Fuente (2011a) suggested that the biconvex cervical vertebra of *Patagoniaemys* is the fourth in the cervical series. Among all non-pleurodiran turtles that have a biconvex cervical vertebra in the middle of the neck, this vertebra is usually the fourth (Williams 1950; Sterli and de la Fuente 2011a: Fig. 11). The only exceptions to this pattern are *Mongolochelys efremovi* (Sukhanov 2000) and some kinosternids (Williams 1950). Furthermore, Sterli and de la Fuente (2011a) believe that the biconcave cervical vertebra of *Patagoniaemys* is the seventh. The presence of a biconcave vertebra in the neck is not common among turtles. Biconcave cervical vertebrae are present in *Mongolochelys efremovi*, where the 6th is the biconcave, and in the eucryptodiras *Ordosemys leios* (Brinkman and Peng 1993) and *Otwayemys cunicularius* (Brinkman and Peng 1993; Gaffney et al. 1998) where the 5th cervical is biconcave. Consequently, Sterli and de la Fuente (2011a) consider the cervical formula of *Patagoniaemys gasparinae* to be (2(, (3(, (4),)5?,?6),)7(,(8). Although a similar cervical formula is recognized in the extant Testudinidae, Emydidae, and Platysternidae, the articulations between cervicals 6, 7, and 8, they are through a double articulation called ginglymoidy (Williams 1950; Sukhanov 2000).

As Gaffney (1985) observed, the presence of a biconvex vertebra in the tail of turtles is generally rare and few exceptions have been found since then. *Patagoniaemys* is one of these exceptions (Sterli and de la Fuente 2011a: Fig. 12). In this taxon the first caudal vertebrae are opisthocoelous and the posterior caudals are procoelous. Another exception is found in the baenid *Boremys pulchra*, where at least caudal centra 4 and 5 are opisthocoelus and the more posterior ones are procoelous (Brinkman and Nicholls 1993).

Gaffney (1996) and Sterli and de la Fuente (2011a) suggested that the characters "formed cervical vertebrae" (character "Cervical articulation A" of Joyce 2007) and "formed caudal vertebrae" appear independently several times during turtle evolution. Following Sterli and de la Fuente (2011a: Fig. 2), "formed cervical vertebrae" appears in the clade that includes *Patagoniaemys gasparinae* and Meiolaniidae, in some baenids, the entire Pleurodira group, and at the base of crown group

Cryptodira. "Formed caudal vertebrae" appears in the clade that includes *Patagoniaemys gasparinae* and Meiolaniidae, in some baenids, in *Xinjiangchelys*, the entire Pleurodira group, and in crown group Cryptodira.

The shape and morphology of the carapace of *Patagoniaemys* are similar to *Mongolochelys efremovi* (see diagnosis and description in Sterli and de la Fuente 2011a). This morphological resemblance is consistent with the phylogenetic analysis performed by Sterli and de la Fuente (2011a) that shows *Patagoniaemys gasparinae, Otwayemys cunicularius*, and *Mongolochelys efremovi* forming a monophyletic, sister group to Meiolaniidae.

6.1.2.3 *Chubutemys* Gaffney, Rich, Vickers-Rich, Constantine, Vacca and Kool 2007

Type species: *Chubutemys copelloi* Gaffney, Rich, Vickers-Rich, Constantine, Vacca and Kool 2007.
Diagnosis: See Gaffney et al. (2007).
Species included: Only *Chubutemys copelloi* Gaffney, Rich, Vickers-Rich, Constantine, Vacca and Kool 2007.

Chubutemys copelloi Gaffney, Rich, Vickers-Rich, Constantine, Vacca and Kool 2007

Holotype: Museo Paleontológico Egidio Feruglio, Trelew, Argentina. MPEF-PV 1236, a partial skull lacking much of the bone from its dorsal surface, ventral surface preserved intact, atlas element disarticulated, and a piece of presumed hyoid in right temporal fossa.
Referred specimens: MPEF-PV 1940, shell fragment with an eighth cervical vertebra; 1941, 1942, and 1943, peripherals.
Diagnosis: See Gaffney et al. (2007).
Locality and age: "Turtle Town," Chubut Province, Argentina. La Paloma Member (not Castaño Member as in Gaffney et al. 2007) of the Cerro Barcino Formation (Aptian–Albian?), Chubut Group (Volkheimer et al. 2009) (Fig. 6.2c, e).
Comments: *Chubutemys copelloi* was named by Gaffney and collaborators in 2007 as a new eucryptodiran turtle characterized by the presence of a processus trochlearis oticus and an enclosed canalis caroticus internus extending to the posterior margin of the pterygoid. In Gaffney's phylogenetic analysis, *Chubutemys* and meiolaniids are related on the basis of the prefrontal–postorbital contact. While recent analyses by Sterli and de la Fuente (2011a, 2012) confirms that *Chubutemys* is closely related to meiolaniids, studies by these authors also show that *Chubutemys* and meiolaniids are stem Testudines and not crown Cryptodira as in Gaffney's hypothesis.

6.2 Discussion

The clade Meiolaniidae is formed by Cenozoic South American and Australasian species of giant land turtles. Although their phylogenetic position among turtles has been discussed extensively in the literature since the nineteenth century

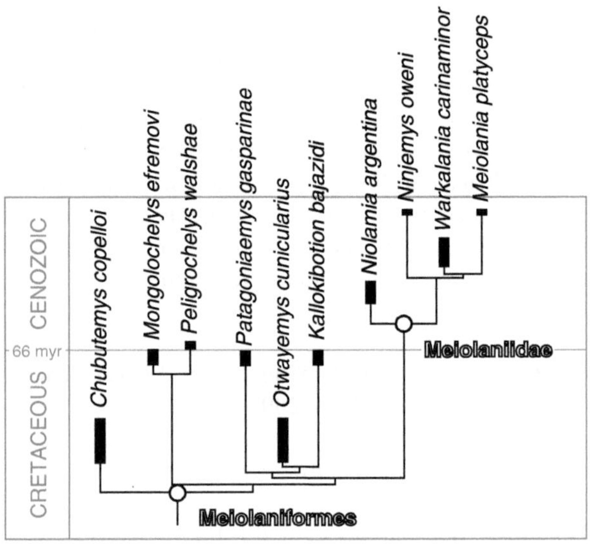

Fig. 6.3 Calibrated phylogeny of Meiolaniformes based on Sterli et al. (2012). Hollow lettering, extinct named clade

(e.g., Boulenger 1887; Baur 1889a, b; Gaffney 1996; Joyce 2007), the internal relationships among meiolaniid taxa are consistent in different cladistics analyses (e.g., Gaffney 1996; Sterli and de la Fuente 2012). Meiolaniids have been associated with Cretaceous forms such as *Mongolochelys efremovi* (Hirayama et al. 2000; Joyce 2007), *Chubutemys copelloi* and *Otwayemys cunicularius* (Gaffney et al. 2007), and more recently also with *Patagoniaemys gasparinae*, *Kallokibotion bajazidi*, and the Palaeocene *Peligrochelys walshae* (Sterli and de la Fuente 2012). The continued association of Meiolaniidae with related taxa in different cladistic analyses prompted Sterli and de la Fuente (2012) to define a new taxon, Meiolaniformes (Fig. 6.3).

Most Meiolaniformes are from Gondwana, with the exception of two Eurasiatic taxa (*Kallokibotion bajazidi* and *Mongolochelys efremovi*). The biogeographic study conducted by Sterli and de la Fuente (2012) suggests that Meiolaniformes originated in present-day South America. The oldest fossils within the clade (*Chubutemys copelloi* and *Otwayemys cunicularius*) indicate a minimum age of origin in the Aptian–Albian (Sterli et al. 2012). The fact that the oldest member of the sister clade is from the Middle Jurassic creates a long ghost lineage subtending Meiolaniformes from the Early Jurassic to the Early Cretaceous (Sterli and de la Fuente 2012). The biogeographic history of Mesozoic Meiolaniformes could be explained by vicariance; however, the Cenozoic history of Meiolaniidae seems to be modeled by the "escalator hopscotch" pattern (Gaffney 1996). The cladograms presented by Gaffney (1996; Fig. 96) and by Sterli and de la Fuente (2012), in which *Niolamia argentina* appears to be the sister group of the other Australasian meiolaniids, are concordant with the breakup of Gondwana components during the Cretaceous. However, to explain the distribution of the meiolaniid species on the islands (e.g., New Caledonia, Lord Howe Island, Vanuatu), Gaffney (1996) suggested the "escalator hopscotch" biogeographic model. In the "escalator hopscotch" model, proposed by McKenna (1983), the biota present on a volcanic island could have originated on an adjacent continent and migrated to the younger island later in time. Consequently, the biota found on the island could be much older than the oldest volcanic rocks comprising the island (McKenna 1983). This interpretation is based on recognition of this pattern on Hawaiian (Mckenna 1983) and Galápagos Islands (Christie et al. 1992).

References

Ameghino F (1899) Sinopsis geológica paleontológica. Suplemento (adiciones y correcciones). Censo Nacional La Plata:1–13

Anderson C (1925) Notes on the extinct chelonian *Meiolania*, with record of a new occurrence. Rec Aust Mus 14:223–242

Baur G (1889a) The systematic position of *Meiolania* Owen. Ann Mag Nat Hist 3:54–62

Baur G (1889b) On *Meiolania* and some points in the osteology of the Testudinata: a reply to Mr G A Boulenger. Ann Mag Nat Hist 4:37–45

Boulenger GA (1887) On the systematic position of the genus *Miolania* Owen (*Ceratochelys* Huxley). Proc Zool Soc Lond 1887:554–555

Boulenger GA (1889) Catalogue of the Chelonians, Rhynchocephalians and Crocodiles in the British Museum (Natural History). New Edition British Museum (Natural History), London

Brinkman DB, Nicholls EL (1993) The skull of *Neurankylus eximius* (Testudines: Baenidae) and a reinterpretation of the relationships of this taxon. J Vertebr Paleontol 13:273–281

Brinkman DB, Peng JH (1993) *Ordosemys leios*, n. gen., n. sp., a new turtle from the Early Cretaceous of the Ordos basin, Inner Mongolia. Can J Earth Sci 30:2128–2138

Christie DM, Duncan RA, McBirney AR, Richards MA, White WM, Harpp KS, Fox CG (1992) Drowned island downstream from the Galapagos hotspot imply extended speciation times. Nature 355:246–248

Gaffney ES (1983) The cranial morphology of the extinct horned turtle, Meiolania platyceps, from the Pleistocene of Lord Howe Island. Australia. Bull Am Mus Nat Hist 175:326–479

Gaffney ES (1985) The cervical and caudal vertebrae of the cryptodiran turtle, *Meiolania platyceps*, from the Pleistocene of Lord Howe Island, Australia. Am Mus Novit 2805:1–29

Gaffney ES (1992) *Ninjemys*, a new name for "*Meiolania*" *oweni* (Woodward), a horned turtle from the Pleistocene of Queensland. Am Mus Novit 3049:1–10

Gaffney ES (1996) The postcranial morphology of *Meiolania platyceps* and a review of the Meiolaniidae. Bull Am Mus Nat Hist 229:1–165

Gaffney ES, Meylan PA, Wyss A (1991) A computer assisted analysis of the relationships of the higher categories of turtles. Cladistics 7:313–335

Gaffney ES, Archer M, White A (1992) *Warkalania*, a new meiolaniid turtle from the Tertiary Riversleigh deposit of Queensland, Australia. Beagle Rec North Territ Mus Arts Sci 9:35–48

Gaffney ES, Kool L, Brinkman DB, Rich TH, Vickers-Rich P (1998) *Otwayemys*, a new cryptodiran turtle from the Early Cretaceous of Australia. Am Mus Novit 3233:1–28

Gaffney ES, Rich TH, Vickers-Rich P, Constantine A, Vacca R, Kool L (2007) *Chubutemys*, a new eucryptodiran turtle from the Early Cretaceous of Argentina, and the relationships of the Meiolaniidae. Am Mus Novit 3599:1–35

Hirayama R, Brinkman DB, Danilov IG (2000) Distribution and biogeography of non-marine Cretaceous turtles. Russ J Herpe 7:181–198

Huxley TH (1887) Preliminary note on the fossil remains of a chelonian reptile, *Ceratochelys sthenurus*, from Lord Howe's Island, Australia. Proc R Soc Lond 42:232–238

Joyce WG (2007) Phylogenetic relationships of Mesozoic turtles. Bull Peabody Mus Nat Hist 48:3–102

Khosatzky LI (1997) Large turtles from the Late Cretaceous of Mongolia. Russ J Herpetol 4:148–154

McKenna MC (1983) Holarctic landmass rearrangement, cosmic event and Cenozoic terrestrial organisms. Ann Mo Bot Gard 70:459–489

Megirian D (1992) *Meiolania brevicollis* sp. nov. (Testudines: Meiolaniidae): a new horned turtle from the Australian Miocene. Alcheringa 16:93–106

Nopcsa F (1923) On the geological importance of the primitive reptilian fauna of the Uppermost Cretaceous of Hungary; with a description of a new tortoise (*Kallokibotion*). Q J Geol Soc 79:100–116

Owen R (1886) Description of some fossil remains of two species of a megalanian genus (*Meiolania*) from "Lord Howe's Island". Phil Trans R Soc Lond 179:181–191

Pascual R, Goin FJ, González P, Ardolino A, Puerta P (2000) A highly derived docodont from the Patagonian Late Cretaceous: evolutionary implications for Gondwanan mammals. Geodiversitas 22:395–414

Pesce A (1979) Estratigrafía del Arroyo Perdido en su tramo medio e inferior, Provincia del Chubut. 7° Congr Geol Arg Neuquén Actas:315–333

Simpson GG (1937) New reptiles from the Eocene of South America. Am Mus Novit 927:1–3

Simpson GG (1938) *Crossochelys*, Eocene horned turtle from Patagonia. Bull Am Mus Nat Hist 74:221–254

Sterli J (2010) Phylogenetic relations among extinct and extant turtles: the position of Pleurodira and the effect of the fossils on rooting crown group turtles. Contrib Zool 79:93–106

Sterli J, de la Fuente MS (2011a) A new turtle from the La Colonia Formation (Campanian–Maastrichtian), Patagonia, Argentina, with remarks on the evolution of the vertebral column in turtles. Palaeontology 54:63–78

Sterli J, de la Fuente MS (2011b) Re-description and evolutionary remarks on the Patagonian horned turtle *Niolamia argentina* Ameghino, 1899 (Testudinata; Meiolaniidae). J Vertebr Paleontol 31:1210–1229

Sterli J, de la Fuente MS (2012) New evidence from the Palaeocene of Patagonia (Argentina) on the evolution and palaeobiogeography of Meiolaniformes (Testudinata, new taxon name). J Syst Palaeontol. doi:10.1080/14772019.2012.708674

Sterli J, de la Fuente MS, Cerda I (2013) A new species of meiolaniform turtle and a revision of the Upper Cretaceous Meiolaniformes of South America. Ameghiniana. Buenos Aires. doi:10.5710/AMGH.x.x.xxxx.582

Sterli J, Pol D, Laurin M (2012) Incorporating phylogenetic incertainty on phylogeny-based on palaeontological dating and the timing of turtle diversification. Cladistics. doi:10.1111/j.1096-0031.2012.00425.x

Sukhanov VB (2000) Mesozoic turtles of Middle and Central Asia. In: Benton MJ, Shishkin MA, Unwin DM, Kurochkin EN (eds) The age of dinosaurs in Russia and Mongolia. Cambridge University Press, Cambridge, pp 309–367

Volkheimer W, Gallego OF, Cabaleri NG, Armella C, Narváez PL, Silva Nieto DG, Páez MA (2009) Stratigraphy, palynology and conchostracans of a Lower Cretaceous sequence at the Cañadón Calcáreo locality, extra-andean central Patagonia: age and paleoenvironmental significance. Cretaceous Res 30:270–282

White AW, Worthy TH, Hawkins S, Bedford S, Spriggs M (2010) Megafaunal meiolaniid horned turtles survived until early human settlement in Vanuatu, Southwest Pacific. Proc Nat Acad Sci USA 107:15512–15516

Williams EE (1950) Variation and selection in the cervical central articulations of living turtles. Bull Am Mus Nat Hist 94:511–561

Woodward AS (1901) On some extinct reptiles from Patagonia, of the genera Miolania, Donilysia, and Genyodectes. Proc Zool Soc Lond 1901:169–184

Chapter 7
Early Differentiation of Mesozoic Turtles

Abstract Panpleurodiran clade Platychelyidae extends from the Oxfordian to the Valanginian of Cuba, South America, and Europe. The oldest member of Platychelyidae was recovered from the Oxfordian of Cuba. The Upper Jurassic record is represented in Europe by only one Kimmeridgian–Tithonian species and in South America by two species. A shallow marine connection between the Western Tethys and Eastern Pacific through central Pangaea was present since the beginning of the Jurassic and acted as a corridor for numerous pelagic reptile groups as the panpleurodiran turtles. Stem turtles (i.e., stem Testudines) from the Late Triassic until the Middle Jurassic have had a Pangaean distribution, and they are known from all continents except Australia and Antarctica. Two species are recognized in South America: the Late Triassic *Palaeochersis talampayensis* and the Middle Jurassic *Condorchelys antiqua*.

Keywords Panpleurodira • Stem Turtles • Australochelyidae • Mesozoic • Pangea

7.1 Early Differentiation of Late Jurassic–Early Cretaceous Panpleurodiran Turtles

7.1.1 Platychelyidae Bräm 1965

In his revision of Solothurn turtle fauna, Bräm (1965) defined the clade Platychelyidae to include the central European *Platychelys oberndorferi* from the Kimmeridgian (Wagner 1853). More recently, Cadena and Gaffney (2005) included all stem Pleurodira (*Platychelys* and *Notoemys* genera) from the Upper Jurassic–Early

M.S. de la Fuente et al., *Origin, Evolution and Biogeographic History of South American Turtles*, Springer Earth System Sciences, DOI 10.1007/978-3-319-00518-8_7,

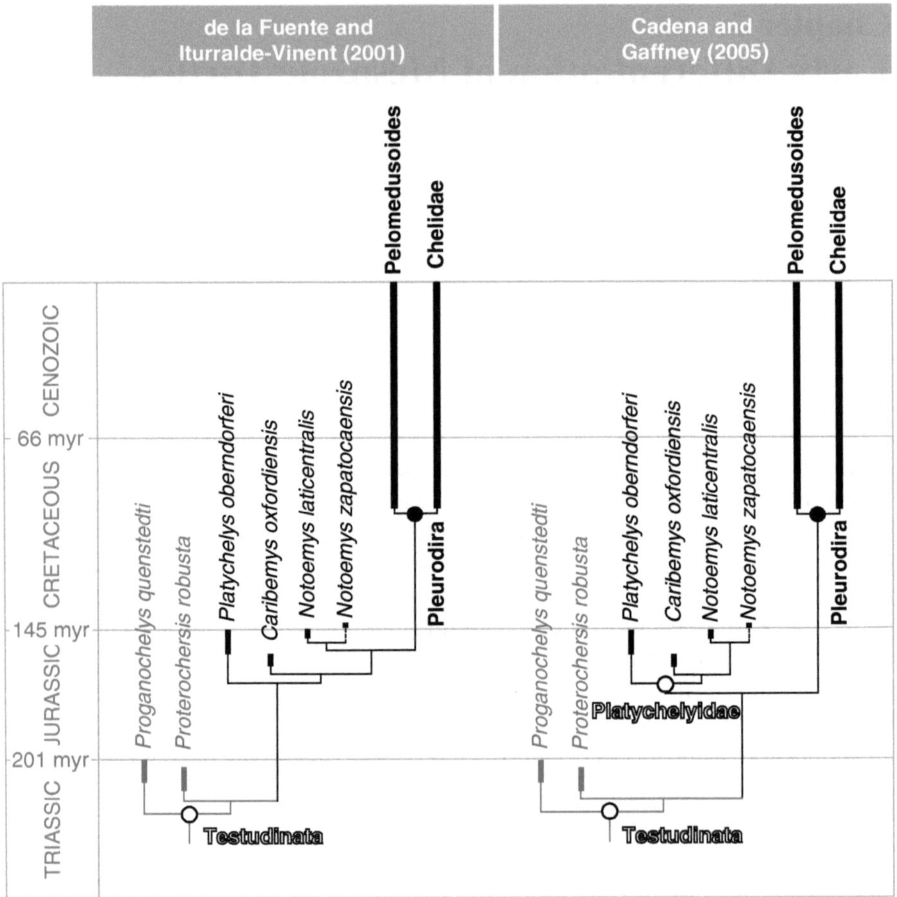

Fig. 7.1 Calibrated phylogenies of competing hypotheses of stem pleurodiran evolution. *White circle*, synapomorphy-based definition of a clade. *Black circle*, crown group

Cretaceous in a monophyletic clade, Platychelyidae, recovered as the sister group of crown Pleurodira. According to Cadena and Gaffney (2005), this clade is characterized by a wide costovertebral tunnel (except *Chelus*), an articulation tubercle on the anterior edge of the first thoracic ribs, and a shell with a wide, straight anterior edge and straight, tapering posterior sides, among other traits. Contrary to the recognition of a monophyletic Platychelyidae, de la Fuente and Iturralde-Vinent (2001) recovered all stem Pleurodira as paraphyletic taxa related to crown Pleurodira. The competing topologies described in these analyses have resulted in different taxonomic definitions and paleobiogeographic interpretations (Fig. 7.1). Here, we follow Cadena and Gaffney (2005).

Genera included: *Caribemys* de la Fuente and Iturralde-Vinent 2001, *Notoemys* Cattoi and Freiberg 1961, and *Platychelys* Wagner 1853.

South American Taxa

7.1.1.1 *Notoemys* Cattoi and Freiberg 1961

Type species: *Notoemys laticentralis* Cattoi and Freiberg 1961.

Species included: *Notoemys laticentralis* Cattoi and Freiberg 1961, *Notoemys zapatocaensis* Cadena and Gaffney 2005. Cadena and Gaffney (2005) *contra* de la Fuente and Iturralde-Vinent (2001) and Lapparent de Broin et al. (2007) included *Caribemys oxfordiensis* de la Fuente and Iturralde-Vinent (2001) in the genus *Notoemys*.

Diagnosis: See Fernández and de la Fuente (1994) and de la Fuente (2007).

Notoemys zapatocaensis Cadena and Gaffney 2005

Holotype: Museo Geológico Ingeominas, Bogotá, Colombia. IPN 15 EAC 140120031, a carapace and posterior part of plastron.

Diagnosis: See Cadena and Gaffney (2005).

Locality and age: El Caucho farm northeast of "Ciudad de Zapatoca," Department of Santander, Colombia. Upper part of the Rosablanca Formation (Guzmán 1985), late Valanginian (Etayo Serna in Cadena and Gaffney 2005) (Fig. 7.2a, b).

Comments: *Notoemys zapatocaensis* was named by Cadena and Gaffney (2005) on the basis of a carapace and a posterior plastron of one specimen. Recently, Cadena and Bloch (2009) and Cadena et al. (2013) described two additional specimens; one of them (a complete articulated shell) preserved complementary traits that provide a more accurate characterization of this species. According to Cadena and Bloch (2009), these include two fairly reduced lateral tuberosities on the margin of the anterior plastral lobe, a shallow notch on the posterolateral margin of the epiplastra, gular scales that are rectangular in shape and much wider than long, a long intergular scale that has a slightly medial contact with the pectorals, resulting in a complete separation of the pectorals that distinguishes *N. zapatocaensis* from *N. laticentralis* and other stem Pleurodira. The phylogenetic analysis addressed by Cadena et al. (2013) as well as the previous one of Cadena and Gaffney (2005) has recovered Platychelyidae as a monophyletic clade. On the contrary, the result obtained by de la Fuente and Iturralde-Vinent (2001) and Joyce (2007) shows *Notoemys* as the sister taxon of crown Pleurodira.

Notoemys laticentralis Cattoi and Freiberg 1961 (Fig. 7.3)

Holotype: Museo Argentino de Ciencias Naturales "Bernardino Rivadavia," Buenos Aires, Argentina. MACN 18043, carapace and anterior plastral lobe.

Diagnosis: See Fernández and de la Fuente (1994) and de la Fuente (2007).

Locality and age: The holotype of *Notoemys laticentralis* was found near Las Lajas, Picunches Department, Neuquén Province, Argentina (de la Fuente 2007: Fig 3.1, locality 7), and the referred specimens were found in Cerro Lotena (de la Fuente 2007: Fig 3.1, locality 14) and Los Catutos (de la Fuente 2007: Fig. 3.1, locality 11), Zapala Department, Neuquén Province, Argentina. These specimens were recovered from the Late Jurassic, lower (Cerro Lotena ?Member)

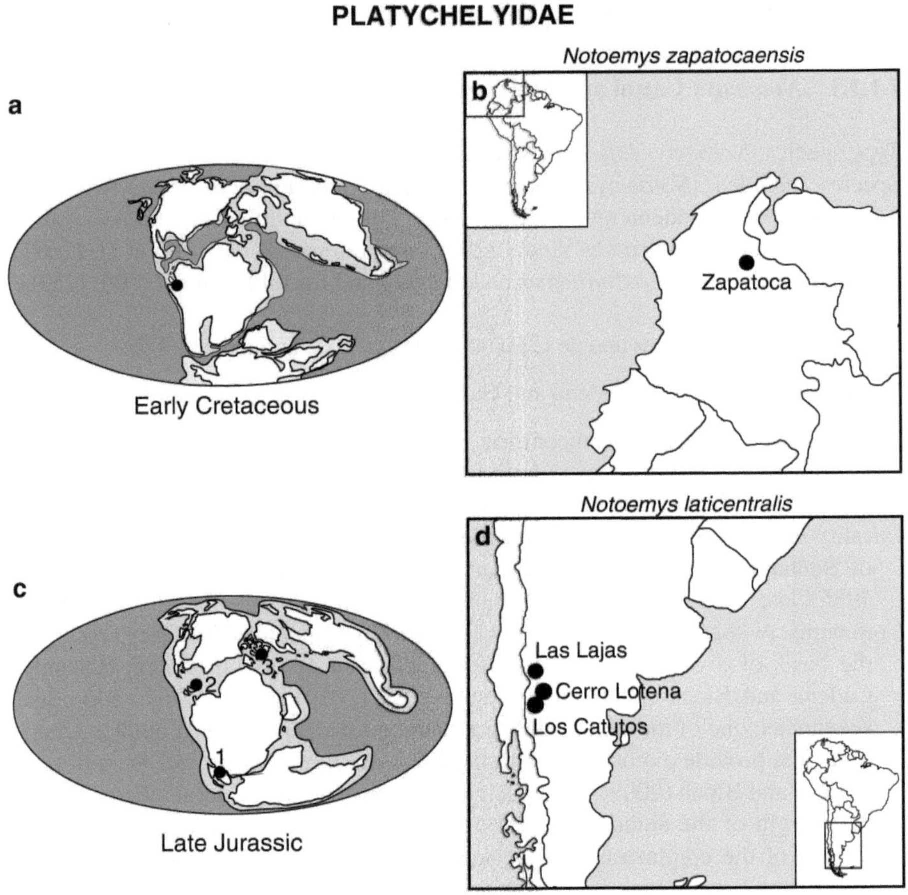

Fig. 7.2 Geographic and stratigraphic distribution of Platychelyidae. (**a**) Paleogeographic reconstruction of continents during the Early Cretaceous. (**b**) Geographic distribution of *Notoemys zapatocaensis* in South America. (**c**) Paleogeographic reconstruction of continents during the Late Jurassic showing the distribution of three platychelyid taxa (1, *Notoemys laticentralis*; 2, *Caribemys oxfordiensis*; 3, *Platychelys oberndorferi*). (**d**) Geographic location of *Notoemys laticentralis*

and middle levels (Las Lajas ?Member) of the Vaca Muerta Formation (lower and middle Tithonian, respectively) (Fig. 7.2c, d).

Comments: *Notoemys laticentralis* was designated as a new taxon of Testudinata by Cattoi and Freiberg (1961) on the basis of the carapace and an unidentified fragment of the plastron. Dr. O. Reig found the specimen in a slate used as construction material in a building in Acassuso, a suburb of Buenos Aires City (Wood and Freiberg 1977). The quarry and level of the slate was identified as Las Lajas, Middle Tithonian of the Vaca Muerta Formation.

Wood and Freiberg (1977) assigned *Notoemys laticentralis* to the clade Plesiochelyidae on the basis of the holotype that was fully described in that paper.

Fig. 7.3 Life reconstruction of *Notoemys laticentralis* from the Late Jurassic of the Vaca Muerta Formation, Argentina

Later, specimens of *Notoemys laticentralis* were reanalyzed and the species was reassigned to Pleurodira (= Panpleurodira sensu Joyce et al. 2004; de la Fuente and Fernández 1989; Fernández and de la Fuente 1993, 1994). Broin and de la Fuente (1993) then coined Notoemydidae to include *N. laticentralis,* and the clade was diagnosed by Fernández and de la Fuente (1994). More recently, Lapparent de Broin et al. (2007) reexamined the skull and cervical vertebrae of the holotype confirming this taxonomic placement. In addition, Lapparent de Broin et al. (2007) showed that the morphology of the sacral area of *Notoemys laticentralis* is essentially pan-pleurodire in its structure. Some of these new results have already been published (Lapparent de Broin and Murelaga 1999; Lapparent de Broin 2000a; de la Fuente and Iturralde-Vinent 2001; de la Fuente 2003; Lapparent de Broin et al. 2007).[1]

7.2 Early Differentiation of Early–Middle Jurassic Turtles

7.2.1 Testudinata Klein 1760

South American Taxa

7.2.1.1 *Condorchelys* Sterli 2008

Type species: *Condorchelys antiqua* Sterli 2008.
Diagnosis: See Sterli (2008).
Species included: Only *Condorchelys antiqua* Sterli 2008.

[1]Courtesy of Indiana University Press.

Fig. 7.4 Life reconstruction of *Condorchelys antiqua* from the Early-Middle Jurassic of the Cañadon Asfalto Formation, Argentina

Condorchelys antiqua Sterli 2008 (Fig. 7.4)

Holotype: Museo Paleontológico Egidio Feruglio, Trelew, Argentina. MPEF-PV 1152, a basicranium.

Diagnosis: See Sterli (2008).

Locality, horizon, and age: The holotype and the referred specimens were recovered at Queso Rallado locality (Rauhut et al. 2002), 5.5 km western from the Cerro Cóndor Village, Chubut Province, Argentina. Queso Rallado is located in outcrops of the Cañadón Asfalto Formation (Stipanicic et al. 1968), which is now considered to be Lower to Middle Jurassic (Tasch and Volkheimer 1970; Nullo and Proserpio 1975; Volkheimer et al. 2009; Cúneo and Bowring 2010) (Fig. 7.5a, b).

Comments: *Condorchelys antiqua* was named and briefly described by Sterli (2008). More recently, a detailed anatomical description of the cranial and postcranial remains of this species was presented by Sterli and de la Fuente (2010). According to Sterli (2008), the significance of *Co. antiqua* lies in two main points: the filling of a gap in the fossil record of continental Jurassic turtles and the position of this species in turtle phylogeny changing the idea about the evolution of some "key" characters. Although the continental Jurassic turtle record is very scarce and fragmentary all the findings are widespread around the world (Europe, India, Central Asia, North America, southern Africa, and southern South America). However, except four species: *Kayentachelys aprix* from the Early Jurassic of North America (Gaffney et al. 1987), *Heckerochelys romani* from the Middle Jurassic of Russia (Sukhanov 2006), *Eileanchelys waldmani* (Anquetin et al. 2009), and *Condorchelys antiqua* (Sterli 2008; Sterli and de la Fuente 2010) where cranial and postcranial remains are articulated or associated, the remaining Early or Middle Jurassic specimens are represented by isolated skulls or shells.

As the main hypotheses about turtle evolution differ in the tempo and the mode of origin of extant clades of turtles, the Jurassic record is important for further testing these hypotheses. Since the first cladistic analyses of turtle evolution

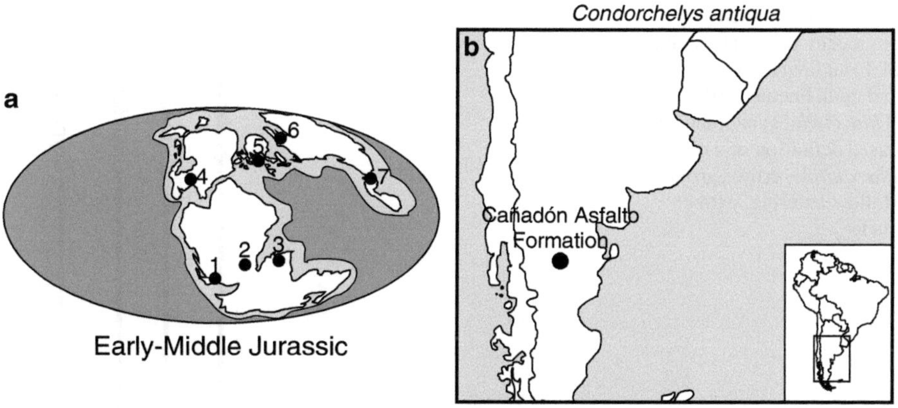

Fig. 7.5 Geographic and stratigraphic distribution of *Condorchelys antiqua*. (**a**) Paleogeographic reconstruction of continents during the Early–Middle Jurassic showing the distribution of extinct turtles (1, *Condorchelys antiqua*; 2, *Australochelys africanus*; 3, *Indochelys spatulata*; 4, *Kayentachelys aprix*; 5, *Eileanchelys waldmani*; 6, *Heckerochelys romani*; 7, *Siamochelys peninsularis*). (**b**) Geographic and stratigraphic location of *Condorchelys antiqua* in South America

performed by Gaffney (1975) and developed by the same author in a series of later contributions (Gaffney et al. 2007, and reference therein), the two living clades of turtles (Cryptodira and Pleurodira) were supported by several synapomorphies. This scheme has not been questioned until the recent papers of Joyce (2007), Sterli and Joyce (2007), Sterli (2008), Anquetin et al. (2009), Sterli (2010), among others. In contrast to Gaffney's hypotheses where all turtles, with the exception of *Proganochelys quenstedti*, *Palaeochersis talampayensis*, and *Australochelys africanus*, are members the two lineages of crown Testudines, the alternative hypothesis (e.g., Joyce 2007; Sterli 2008) suggests that many extinct turtles previously known to be crown turtles are instead stem-group turtles (Fig. 7.6). These different hypotheses suggest a distinct timing of crown-group turtle origin. According to Gaffney's hypothesis, the origin of the crown-group Testudines occurred during the Upper Triassic, whereas the alternative hypothesis suggests that the origin of the crown-group turtles was more recent. *Kayentachelys aprix* was originally proposed as the oldest pancryptodiran turtle by Gaffney et al. (1987) because of the presence of a processus trochlearis oticum, among other characteristics. However, Sterli and Joyce (2007) pointed out that the anterior wall of the otic region of *K. aprix* is thicker than in *Proganochelys quenstedti* but does not develop a processus trochlearis oticum (see additional discussion in Gaffney and Jenkins 2010 and reply in Joyce and Sterli 2012). The same morphology present in *K. aprix* is also found in *Co. antiqua*. The cladistic analyses performed by Sterli (2008) generally support the topology obtained by Joyce (2007) and suggest that this Jurassic taxon from the Jurassic of Patagonia is phylogenetically closer to *K. aprix*. Consequently, the discovery of new fossil turtle taxa from the Triassic and Jurassic gives new information about the first steps in turtle evolution, and it allows us to reinterpret some characters and to test the competing hypotheses about turtle evolution.

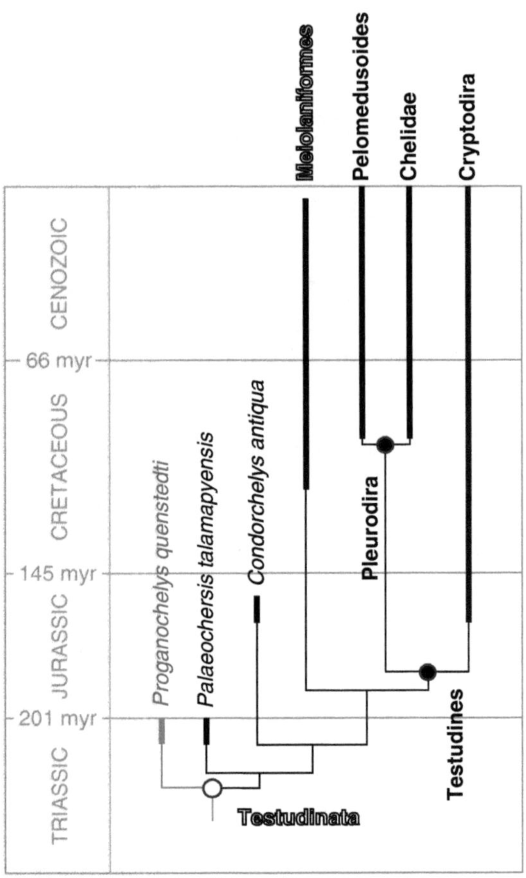

Fig. 7.6 Calibrated phylogeny of the main clades of Testudinata based on Sterli and de la Fuente (2012). *White circle*, synapomorphy-based definition of a clade. *Black circle*, crown group. Hollow lettering, extinct clade

Recent papers on continental Early and Middle Jurassic turtles (e.g., Sukhanov 2006; Anquetin et al. 2009; Anquetin 2010; Sterli and de la Fuente 2010) have shown that they also lack specialized jaw closure mechanism (the absence of processus trochlearis). This available evidence suggests that the origin of the pulley systems in turtles is at least 60 Myr younger than previously thought (Gaffney et al. 1987; Gaffney and Jenkins 2010).

7.3 Early Differentiation of Late Triassic–Early Jurassic Turtles

7.3.1 Australochelyidae Gaffney and Kitching 1994

Gaffney and Kitching (1994) coined Australochelyidae based on *Australochelys africanus* holotype (a skull) which was discovered at Bormansdrift, Orange Free

State, South Africa, from *Tritylodon* Acme Zone (middle Elliot Formation, Early Jurassic). Lately, a detailed description of this species was presented by Gaffney and Kitching (1995). Rougier et al. (1995) named *Palaeochersis talampayensis* from Upper Triassic of La Rioja (Argentina) and suggested this new taxon belongs to Australochelyidae. According to Rougier et al. (1995) and Sterli et al. (2007), the clade Australochelyidae is characterized by the roofing of the posterodorsal margin of the temporal fossa by an overhanging process of the skull roof and by the depressions in the wide transverse occipital plane for the nuchal musculature. Other authors (Hirayama et al. 2000; Gaffney et al. 2007; Joyce 2007; Cadena et al. 2012) have not recovered the Australochelyidae as a monophyletic clade.

Genera included: *Australochelys* Gaffney and Kitching 1994 and *Palaeochersis* Rougier, de la Fuente, and Arcucci 1995.

South American Taxa

7.3.1.1 *Palaeochersis* Rougier, de la Fuente, and Arcucci 1995

Type species: *Palaeochersis talampayensis* Rougier, de la Fuente, and Arcucci 1995.
Diagnosis: See Rougier et al. (1995) and Sterli et al. (2007).
Species included: Only *Paleochersis talampayensis* Rougier, de la Fuente, and Arcucci 1995.

Palaeochersis talampayensis Rougier, de la Fuente, and Arcucci 1995 (Fig. 7.7)

Holotype: Universidad Nacional de La Rioja, La Rioja Province, Argentina. PULR 68, skull with lower jaw, cervical, thoracic, and caudal vertebrae, both fore and hind limbs, shoulder and pelvic girdles, and carapace and plastron.
Diagnosis: See Rougier et al., de la Fuente, Arcucci (1995) and Sterli et al. (2007).
Locality, horizon, and age: 2 km northwest from La Esquina, La Rioja Province, Argentina. Upper section of the Los Colorados Formation, referred to the Middle Norian (Upper Late Triassic) based on magnetostratigraphic studies (Santi Malnis et al. 2011) (Fig. 7.8a, b).
Comments: The scenario presented by Sterli et al. (2007) confirms in general terms the hypothesis proposed by Rougier et al. (1995) in which *Palaeochersis talampayensis* and *Australochelys africanus* form a monophyletic group, named Australochelyidae, and that they are the sister group of the remaining turtles except *Proganochelys quenstedti*. Australochelyidae is supported by two synapomorphies: wide transverse occipital plane with depressions for the nuchal musculature and the posterodorsal margin of the temporal fossa roofed by an overhanging process of the skull roof. In this sense, the phylogenetic scenario proposed by Sterli et al. (2007) is also different to that proposed by Joyce (2007). The preferred phylogenetic analysis of Joyce (2007) shows that *Palaeochersis talampayensis* together with *Australochelys africanus*, and the remaining turtles (except *Proganochelys quenstedti*), form a trichotomy. Regarding Joyce (2007), this trichotomy is due to the poor information available at that moment for

Fig. 7.7 Life reconstruction of *Palaeochersis talampayensis* from the Late Triassic of the Los Colorados Formation, Argentina

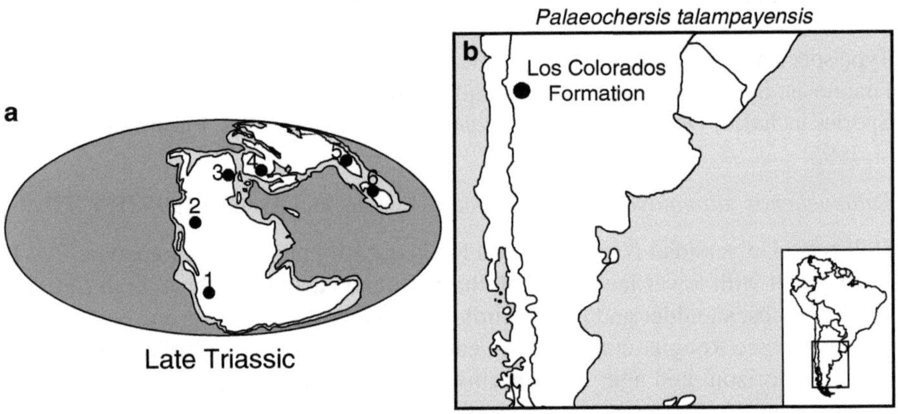

Fig. 7.8 Geographic and stratigraphic distribution of Late Triassic stem Testudines. (**a**) Paleogeographic reconstruction of continents during the Late Triassic showing the distribution of extinct turtles (1, *Palaeochersis talampayensis*; 2, *Chinlechelys tenertesta*; 3, *Proganochelys* sp.; 4, *Proganochelys quenstedti and Proterochersis robusta*; 5, *Odontochelys semitestacea*; 6, *Proganochelys ruchae*). (**b**) Geographic and stratigraphic location of *Palaeochersis talampayensis* in South America

Palaeochersis talampayensis. However, the detailed description of *Palaeochersis talampayensis* presented by Sterli et al. (2007), together with its phylogenetic relationships, shows that both genera (*Australochelys africanus* and *Palaeochersis talampayensis*) share derived characteristic that allows us to group them together under Australochelyidae.

The position of *Palaeochersis talampayensis* as an Australochelyidae, as was proposed by Rougier et al. (1995), is challenged by Lapparent de Broin's ideas. In different papers, this author has proposed that *Palaeochersis talampayensis* is possible a "pre-pleurodiran turtle" (Lapparent de Broin 2000a, b) or even a "true basic pleurodiran" together with *Proterochersis robusta* (Lapparent de Broin et al. 2004). Her pleurodiran hypothesis was primarily based on the suture of the pelvis with the carapace with a beginning of a modification of the surrounding areas in *Palaeochersis talampayensis*. However in *Palaeochersis talampayensis* specimens, the skull and neck does not show any pleurodiran feature. On the contrary, the skull and neck of *Palaeochersis talampayensis* show basal traits like the presence of lacrimal and supratemporal bones, the presence of a quadrate pocket, a foramen trigemini partially enclosed, middle ear limits partially developed, presence of an interpterygoid vacuity, presence of a cultriform process, and a high dorsum sellae. Unfortunately, neither the skull nor the neck is preserved in *Proterochersis robusta,* and the skull is absent in *Platychelys oberndorferi,* so the comparison with this basal panpleurodiran turtles is not possible.

Lapparent de Broin et al. (2004: Fig. 3) proposed an evolutionary scheme to fix the "evolutionary grade" of the pleurodiran clade Dortokidae (Upper Cretaceous of Europe) and included *Palaeochersis talampayensis* and *Proterochersis robusta* at the base of the pleurodiran "grade." This "evolutionary grade" is recognized by Lapparent de Broin et al. (2004) in both Triassic taxa characterized by "sutured pelvis and unretractile neck." The additional characters shared by *Palaeochersis talampayensis* and *Proterochersis robusta*, such as the partly rounded carapace, without anterior notch and the posterior semirounded notch placed on the pygal zone, were also pointed out by Lapparent de Broin et al. (2004) as indicative of their pleurodiran affinities. To compare Rougier et al. (1995) and Sterli et al. (2007) hypothesis with Lapparent de Broin's idea, Sterli et al. (2007) performed a cladistic analysis using constraints. The authors constrained the position of *Palaeochersis talampayensis* within the panpleurodiran turtles and this analysis resulted in trees 29 steps longer that the two most parsimonious trees found in this analysis (Sterli et al. 2007). As a consequence, and contrary to the hypothesis proposed by Lapparent de Broin in different papers (Lapparent de Broin 2000a, b; Lapparent de Broin et al. 2004), the most parsimonious result in a cladistic context is that *Palaeochersis talampayensis* together with *Australochelys africanus* form a monophyletic assemblage, being *Palaeochersis talampayensis* a stem Testudines, not a Panpleurodira. The pelvis–shell attachment by suture is a trait that has evolved, at least twice during turtle evolution: in *Palaeochersis talampayensis* (and probably in *Australochelys africanus*) and in panpleurodiran turtles. As a result of Sterli et al. (2007) cladistic analysis, the pelvis–shell relationship found in *Palaeochersis talampayensis* would not be homologous to that present in *Proterochersis robusta* and other Panpleurodira. Besides the attachment by suture of the ischium with the xiphyplastron in *Palaeochersis talampayensis* is not made by the lateral ischial process, as it is seen panpleurodires, but made instead through its medial surface, not including its well-developed lateral process. This trait would be correlated with the narrow posterior plastral lobe of *Palaeochersis talampayensis* that ends in a rounded margin (without the anal notch present in *Proterochersis robusta* and remaining panpleurodires).[2]

[2]Courtesy of Schweizerbart Science Publishers (www.schweizerbart.de).

7.4 Discussion

7.4.1 *Late Jurassic–Early Cretaceous Panpleurodiran Paleobiogeography*

Platychelyidae (sensu Cadena and Gaffney 2005) extends from the Oxfordian (*N. oxfordiensis*) to the Valanginian (*N. zapatocaensis*). The oldest member of Platychelyidae, *Caribemys oxfordiensis,* was recovered from the Oxfordian Jagua Formation (near Vignales, Cuba). The European Upper Jurassic record is represented by *Platychelys oberndorferi* (from the Kimmeridgian of Solothurn, Switzerland, and the Tithonian of Kelheim, Germany; Bräm 1965) and aff. *Platychelys* (from the Kimmeridgian of Guimarota, Portugal; Broin 1988; Lapparent de Broin 2001). The Upper Jurassic record of South America is completed by the Tithonian *Notoemys laticentralis* (Wood and Freiberg 1977; de la Fuente and Fernández 1989; Fernández and de la Fuente 1993, 1994; de la Fuente 2007, Lapparent et al. 2007). The youngest record of Platychelyidae comes from the Valanginian of Colombia (South America).

Although the geographical range for this clade appears to be extensive (Cadena and Gaffney 2005), paleogeographic reconstruction of the Late Jurassic (Smith and Briden 1977) shows a much closer proximity of the mentioned localities. In this sense, a shallow marine connection between the Western Tethys and Eastern Pacific through central Pangaea (= "Hispanic Corridor" of Smith 1983) was present at the beginning of the Jurassic. Several authors have recognized the same genera and even species of marine invertebrates in both geographic realms (Damborenea and Manceñido 1979; Hallam 1983; Riccardi 1991; Ballent and Wahtley 2000; Damborenea 2000; Aberham 2001). Likewise, these realms share numerous genera of marine reptiles that also date to at least the Bajocian (Gasparini and Iturralde-Vinent 2006 and reference therein). According to Gasparini (e.g., Gasparini 1978, 1992; Gasparini and Fernández 1996, 2005), the possibility that this Caribbean Seaway acted as a connection cannot be ruled out. In addition, as argued by Iturralde-Vinent (2003, 2004), the opening of this oceanic seaway represents an important paleooceanographic event, linking the Tethys, Central Atlantic, and Eastern Pacific oceans since the Oxfordian and until the uplift of Central America.

7.4.2 *Late Triassic to Middle Jurassic Stem Turtles*

Stem turtles (i.e., stem Testudines) from the Late Triassic until the Middle Jurassic have had a Pangaean distribution, and they are known from all continents except Australia and Antarctica. The knowledge of Early to Middle Jurassic stem turtles has been scarce until the twenty-first century. Only *Australochelys africanus* and *Kayentachelys aprix* were known before 2000 (Gaffney et al. 1987; Gaffney and Kitching 1994, 1995). After that, several discoveries have been filling the gap between the oldest known turtles from the Upper Triassic and the Upper Jurassic turtles known from Europe, North America, and Asia. Nowadays there are at least six species from Early–Middle Jurassic located outside Testudines (e.g., *Condorchelys antiqua*, *Indochelys spatulata*, *Kayentachelys aprix*, *Australochelys africanus*, *Heckerochelys romani*, *Eileanchelys waldmani*), and only two of them inhabited future Gondwana landmasses (*Co. antiqua* and *I. spatulata*). Because of the restricted findings of extinct turtles from the Southern Hemisphere continents during that time, *Co. antiqua* and *I. spatulata* bring valuable information to understand the evolution of certain structures and characters and to inquiry into the origin of crown-group Testudines.

The history of Triassic turtle findings is much older and more complex than for Early to Middle Jurassic turtles. Triassic turtles are known since 1884, thanks to a variety of German specimens studied in the nineteenth century by Baur (1887), Zakrszeswski (1888), Quenstedt (1889), Fraas (1899, 1913), and Jaekel (1914, 1918a, b) and restudied during the twentieth century by Parsons and Williams (1961), Gaffney and Meeker (1983), and Gaffney (1985, 1990). All the specimens are Norian in age and were collected in a few spotty quarries throughout Germany. The European material has been referred to two monospecific genera: *Proganochelys quenstedti* and *Proterochersis robusta*. *Proganochelys quenstedti* is reported as the sister group of all the remaining turtles (with the exception of the recently discovered *Odontochelys semitestacea*) (Gaffney and Meeker 1983), while *Proterochersis robusta* was considered the earliest known pleurodire (Fraas 1913; Gaffney and Meylan 1988; Sterli et al. 2007) or stem Testudines (Rougier et al. 1995; Joyce 2007; Sterli 2008, 2009, 2010; Anquetin et al. 2009).

Other Triassic turtles whose remains have been assigned to Proganochelidae are known from Thailand and Greenland. The Thai specimen comes from the Huai Hin Lat Formation and has been described by Broin (1984). The very fragmentary nature of the Thai Triassic turtle add not significant anatomical information for evaluating the evolution of the basal turtles, but it is showing the widespread distribution of these basal turtles. Fragmentary remains of Triassic turtles in the Fleming Fjord Formation (Greenland) have been reported by Jenkins and coauthors (Jenkins et al. 1994). The specimens are attributed to *Proganochelys* sp.

More recently, two significant discoveries shed light on the evolution, paleobiogeography, and palaeocology of basal turtles. A new thin-shelled extinct turtle *Chinlechelys tenertesta* Joyce, Lucas, Scheyer, Heckert and Hunt 2009 from the Upper Triassic (Norian) Chinle group of New Mexico (North America), and the most basal member of Testudinata *Odontochelys semitestacea* Li, Wu, Rieppel, Wang and Zhao 2008 from Upper Triassic (lower Carnian) Nampanjian Trough Basin, near Guanling, Guizhou Province (China), have been described. Particularly, the Chinese species has yielded significant data to improve the knowledge concerning habitat preferences of basal members of Testudinata and on the origin of the turtle body plan (Li et al. 2008; Nagashima et al. 2009, 2012; Rieppel 2012). Joyce and Gauthier's (2004) analysis of habitat preferences of extant and extinct turtles based on forelimb measurements suggested that *Proganochelys quenstedti* and *Palaeochersis talampayensis* were well adapted to terrestrial habitats. Prior to the discovery of *Odontochelys semitestacea*, Joyce and Gauthier's (2004: Fig. 3) optimization of aquatic versus terrestrial habitats in turtles showed that the last common ancestor of crown turtles lived in fresh water. The relatively short anterior autopodium of the two successive outgroups of crown Testudines (*Proganochelys quenstedti* and *Palaeochersis talampayensis*) suggests they were terrestrial. Joyce and Gauthier's (2004) study suggests that the habitat preference of stem turtles as well as the origin of the turtles likely took place in a terrestrial environment more than in an aquatic one. This assumption seems not to be correct if the aquatic habit of *Odontochelys semitestacea* is corroborated (Li et al. 2008).

With the exception of *Palaeochersis talampayensis*, from the Upper Triassic of Argentina, all the abovementioned Triassic turtles are known from Laurasia landmasses. In Gondwana, the oldest record of Testudinata outside Argentina is *Australochelys africanus* (Gaffney and Kitching 1994; Lapparent de Broin 2000b) from the Early Jurassic of South Africa. The fossil record suggests turtles have been widespread in Pangaea (Europe, North America, Asia, South America, and Africa) soon after their appearance. Turtles with Pangaean distribution were common until the Middle to Late Jurassic. After the fragmentation of Pangaea, several groups of turtles (e.g., xinjiangchelyids, sinemydids, pleurosternids) started to show a restricted distribution in different regions of Laurasia. Unfortunately, the continental record of Gondwanan turtles in the Upper Jurassic is almost inexistent; only one species is known from Uruguay, but it is still undescribed (Perea et al. 2012). We believe new findings will bring more information about the early evolution of continental turtles in Gondwana.

References

Aberham M (2001) Bivalve paleobiogeography and the Hispanic Corridor: time of opening and the effectiveness of a proto-Atlantic seaway. Palaeogeogr Palaeoclimatol Palaeoecol 165:375–394

Anquetin J (2010) The anatomy of the basal turtle *Eileanchelys waldmani* from the Middle Jurassic of the Isle of Skye, Scotland. Earth Environ Trans R Soc Edinb 101:67–96

Anquetin J, Barrett PM, Jones MEH, Moore-Fay S, Evans SE (2009) A new stem turtle from the Middle Jurassic of Scotland: new insights into the evolution and palaeoecology of basal turtles. Proc R Soc B 276:879–886

Ballent S, Wahtley R (2000) The distribution of the Mesozoic ostracod genus *Procytherura* Wahtley: palaeogeographical implications with special reference to Argentina. Alcheringa 24:229–242

Baur G (1887) Ueber den Ursprung der Extremitaten der Ichthyopterygia. Jahrbr Mitt Oberrhein Geol Verein 20:17–20

Bräm H (1965) Die Schildkröten aus dem oberen Jura(Malm) der Gegend von Solothurn. Schweizer Paläont Abhandl 83:1–190

Cadena EA, Bloch JI (2009) Two new specimens of the platychelyid turtle *Notoemys zapatocaensis* from the Early Cretaceous (Valanginian) of Colombia. Spec Publ R Tyrrell Mus Turt Symp Abstr Programs 1:29–31

Cadena E, Gaffney ES (2005) *Notoemys zapatocaensis*, a new side-necked turtle (Pleurodura: Platychelyidae) from the Early Cretaceous of Colombia. Am Mus Novit 3470:1–19

Cadena EA, Ksepka DT, Jaramillo CA, Bloch JI (2012) New pelomedusoid turtles from the Late Palaeocene Cerrejón Formation of Colombia and their implications for phylogeny and body size evolution. J Syst Palaeontol 10:313–331

Cadena EA, Jaramillo CA, Bloch JI (2013) New Material of the Platychelyid Turtle Notoemys zapatocaensis from the Early Cretaceous of Colombia; Implications for Understanding Pleurodira Evolution. In: Brinkman DB, Holroyd PA, Gardner JD (eds) Morphology and evolution of turtles, vol 8, pp 105–120. Proceedings of Gaffney Turtle symposium (2009) in Honor of Eugene Gaffney, Vertebrate Paleobiology and Paleoanthropology. Springer, Dordrecht

Cattoi N, Freiberg M (1961) Nuevo hallazgo de Chelonia extinguidos en la República Argentina. Physis 22:202

Cúneo R, Bowring S (2010) Dataciones geocronológicas preliminares en la Cuenca Cañadón Asfalto, Jurásico de Chubut, Argentina. Implicancias geológicas y paleontológicas. Abstracts 10° Congr Arg de Paleontol Bioestr 7° Congr Latinoamer Paleontol 153. La Plata

Damborenea S (2000) Hispanic Corridor: its evolution and the biogeography of bivalve molluscs. GeoRes Forum 6:369–380

Damborenea S, Manceñido M (1979) On the paleogeography distribution of the pectinid genus *Weyla* (Bivalvia, Lower Jurassic). Paleogeogr Palaeoclimatol Palaeoecol 27:85–102

de Broin F (1984) *Proganochelys ruchae* n. sp., chélonien du Trias supérieur de Thailande. Stud Palaeochel 1:87–97

de Broin F (1988) Les Tortues et le Gondwana. Examen des rapport entre le fractionemment du Gondwana et la dispersión géographique dess tortues pleurodires àpartir de Crétacé. Stud Geol Salmant Studia Palaeochel 2:103–142

de Broin F, de la Fuente MS (1993) Les tortues fossiles d'Argentine: synthèse. Ann Paléontol 79:169–232

de Gasparini ZB (1978) Consideraciones sobre los Metriorhynchidae (Crocodilia, Mesosuchia): su origen, taxonomía y distribución geográfica. Obra Cent Mus La Plata 5:1–9

de Gasparini ZB (1992) Marine reptiles. In: Westermann G (ed) The Jurassic of the Circum-Pacific. Cambridge University Press, New York

de Gasparini ZB, Fernández M (1996) Biogeographical affinities of the Jurassic marine reptiles of South America. In: Riccardi A (ed) Advances in Jurassic research, GeoResearch forum. Transtec Publications, Zurich

de Gasparini ZB, Fernández M (2005) Jurassic marine reptiles in the Neuquen Basin. In: Veiga GD, Spalletti LA, Howell JA, Schwarz E (eds) The Neuquén Basin, Argentina: a case study in sequence stratigraphy and basin dynamics, vol 252, Geological society special publication. Geological Society, London, pp 279–294

de Gasparini ZB, Iturralde-Vinent M (2006) The Cuban Oxfordian herpetofauna in the Caribbean Seaway. N Jb Geol Paläont Abh 240:343–371

de la Fuente MS (2003) Two new Pleurodiran turtles from the Portezuelo Formation (Upper Cretaceous) of northern Patagonia, Argentina. J Paleontol 77:559–575

de la Fuente MS (2007) Testudines. In: de Gasparini ZB, Coria R, Salgado L (eds) Patagonian Mesozoic Reptiles. Indiana University Press, Bloomington

de la Fuente MS, Fernández MS (1989) *Notoemys laticentralis* Cattoi & Freiberg, 1961, from the Upper Jurassic of Argentina: a member of the Infraorder Pleurodira (Cope, 1868). Stud Geol Salmant Stud Palaeochel 3:25–32

de la Fuente MS, Iturralde-Vinent M (2001) A new pleurodiran turtle from the Jagua Formation (Oxfordian) of western Cuba. J Paleontol 75:860–869

Fernández MS, de la Fuente MS (1993) Las tortugas casiquélidias de las calizas litográficas titonianas del área Los Catutos Argentina. Ameghiniana 30:283–295

Fernández MS, de la Fuente MS (1994) Redescription and phylogenetic position of *Notoemys*: The oldest Gondwanian pleurodiran turtle. N Jahrb Geol Paläont Abh 19:81–105

Fraas E (1899) *Proganochelys quenstedti* Baur (*Psamobates keuperina* Qu.) Ein neuer Fund der Keuperschildkrote aus dem Stubensandstein. Jahreshberg Verein Vaterl Naturk Württemberg 60:401–424

Fraas E (1913) *Proterochersis*, eine pleurodire Schildkrote aus dem keuper. Jahreshberg Verein Vaterl Naturk Württemberg 80:13–30

Gaffney ES (1975) A phylogeny and classification of the higher categories of turtles. Bull Am Mus Nat Hist 155:387–436

Gaffney ES (1985) The shell morphology of the Triassic turtle. Proganochelys N Jb Geol Paläeontol Abh 170:1–26

Gaffney ES (1990) The comparative osteology of the Triassic turtle *Proganochelys*. Bull Am Mus Nat Hist 194:1–263

Gaffney ES, Jenkins FA (2010) The cranial morphology of *Kayentachelys*, an Early Jurassic cryptodire, and the early history of turtles. Acta Zool 11:35–68

Gaffney ES, Kitching JW (1994) The most ancient African turtle. Nature 369:55–58

Gaffney ES, Kitching JW (1995) The morphology and relationships of *Australochelys*, an Early Jurassic Turtle from South Africa. Am Mus Novit 3130:1–29

Gaffney ES, Meeker LJ (1983) Skull morphology of the oldest turtle: a preliminary description of *Proganochelys quenstedti*. J Vertebr Paleontol 3:25–28

Gaffney ES, Meylan P (1988) A phylogeny of turtles. In: Benton MJ (ed) The phylogeny and classification of the tetrapods 1: amphibians, reptiles, birds, vol 35A, Systematics association, special volume. Clarendon, Oxford

Gaffney ES, Hutchison JH, Jenkins FA, Meeker LJ (1987) Modern turtle origins: the oldest known cryptodire. Science 237:289–292

Gaffney ES, Rich TH, Vickers-Rich P, Constantine A, Vacca R, Kool L (2007) *Chubutemys*, a new eucryptodiran turtle from the Early Cretaceous of Argentina, and the relationships of the Meiolaniidae. Am Mus Novit 3599:1–35

Guzmán G (1985) Los Grifeidos infracretácicos *Aetostreon couloni* y *Ceratrostreon boussingaulti*, de la Formación Rosablanca, como indicadores de oscilaciones marinas. In: Etayo S, Proyecto Cretácico Publ Esp Ing Bogotá 12: 1–16

Hallam A (1983) Early and Mid-Jurassic molluscan biogeography and the establishment of the central Atlantic seaway. Palaeogeogr Palaeoclimatol Palaeoecol 43:181–193

Hirayama R, Brinkman DB, Danilov IG (2000) Distribution and biogeography of non-marine Cretaceous turtles. Russ J Herpetol 7:181–198

Iturralde-Vinent M (2003) A brief account of the evolution of the Caribbean seaway: Jurassic to Present. In: Prothero D, Ivany L, Nesbitt E (eds) From greenhouse to icehouse: the Marine Eocene-Oligocene Transition. Columbia University Press, New York

Iturralde-Vinent M (2004) The conflicting paleontologic vs stratigraphic record of the formation of the Caribbean Seaway. In: Bartolini C, Buffer R, Blickwede J (eds) The Circum-Gulf of Mexico and the Caribbean: hydrocarbon habitat, basin formation, and plate tectonic. Mein Amer Assoc Petrol Geol, vol 79, pp 75–88

Jaekel O (1914) Ueber die Wirbeltierfunde in der oberen Trias von Halberstadt. Paläontol Z 1:155–215

Jaekel O (1918a) Die Wirbeltierfunde aus dem Keuper von Halberstadt. Serie II. Testudinata. Teil 1. *Stegochelys dux* n g., n. sp. Paläontol Z 2:88–214

Jaekel O (1918b) [Untitled note] Name *Stegochelys* changed to *Triassochelys*. Paläontol Z 2:251

Jenkins FA Jr, Shubin NH, Amaral WW, Gatesy SM, Schaff CR, Clemmensen LB, Downs WR, Davidson AR, Bonde N, Osbaeck F (1994) Late Triassic continental vertebrates and depositional environments of the Fleming Fjord Formation, Jameson Land, East Greenland. Medd Gron Geosci 32:3–25

Joyce WG (2007) Phylogenetic relationships of Mesozoic turtles. Bull Yale Peabody Mus 48:3–102

Joyce WG, Gauthier JA (2004) Palaeoecology of the Triassic stem turtles sheds new light on turtle origins. Proc R Soc B 271:1–5

Joyce WG, Sterli J (2012) Congruence, non-homology, and the phylogeny of basal turtles. Acta Zool 93:149–159

Joyce WG, Parham JF, Gauthier JA (2004) Developing a protocol for the conversion of rank-based taxon names to phylogenetically defined clade names, as exemplified by turtles. J Paleontol 78:989–1013

Joyce WG, Lucas SG, Scheyer TM, Heckert AB, Hunt AP (2009) A thin-shelled reptile from the Late Triassic of North America and the origin of the turtle shell. Proc R Soc B 276:507–513

Lapparent de Broin F (2000a) The oldest pre-Podocnemidid turtle (Chelonii, Pleurodira), from the Early Cretaceous, Ceará state, Brazil, and its environment. Treb Mus Geol Barcelona 9:43–95

Lapparent de Broin F (2000b) African chelonians from the Jurassic to the present: phases of development and preliminary catalogue of the fossil record. Palaeontol Africana 36:43–82

Lapparent de Broin F (2001) The European turtle fauna from the Triassic to the Present. Dumerilia 4:155–217

Lapparent de Broin F, Murelaga X (1999) Turtles from the upper cretaceous of Laño (Iberian Peninsula). In: Asribia H et al. Geology and paleontology of the upper Cretaceous vertebrate-bearing beds of the Laño Quarry (Basque-Cantabrian region, Iberian Peninsula. Estud Mus Cienc Nat Alava 14:135–211

Lapparent de Broin F, Murelaga Bereikua X, Codrea V (2004) Presence of Dortokidae (Chelonii, Pleurodira) in the Earliest Tertiary of Jibou Formation, Romania: Paleobiogeographical implications. Acta Palaeontol Romaniae 4:203–215

Lapparent de Broin F, de la Fuente MS, Fernández MS (2007) *Notoemys* (Chelonii, Pleurodira), Late Jurassic of Argentina: new examination of the anatomical structures and comparisons. Rev Paléobiol 26:99–136

Li C, Wu X, Rieppel O, Wang L, Zhao L (2008) An ancestral turtle from the Late Triassic of south-western China. Nature 45:497–501

Nagashima H, Sugahara F, Takechi M, Ericsson R, Kawashima-Ohya Y, Narita Y (2009) Evolution of the turtle body plan by the folding and creation of new muscle connections. Science 325:193–196

Nagashima H, Kuraku S, Uchida K, Kawashima-Ohya Y, Narita Y, Kuratani S (2012) Origin of the turtle body plan: the folding theory to illustrate turtle specific developmental repatterning. In: Brinkman DB, Holroyd PA, Gardner JD (eds) Morphology and evolution of turtles, vol 4, pp 37–50. Proceedings of Gaffney Turtle Symposium (2009) in Honor of Eugene Gaffney. Vertebrate Paleobiology and Paleoanthropology, Springer, Dordrecht

Nullo F, Proserpio C (1975) La Formación Traquetrén en Cañadón del Zaino (Chubut) y sus relaciones estratigráficas en el ámbito de la Patagonia, de acuerdo a la flora, República Argentina. Rev Asoc Geol Arg 30:133–150

Parsons TS, Williams EE (1961) Two Jurassic turtles skulls: a morphological study. Bull Mus Comp Zool 125:43–107

Perea D, Soto M, Sterli J, Mesa V, Toriño P, Roland G, Da Silva J (2012) A new Late Jurassic-?earliest Cretaceous continental turtle from Western Gondwana (Tacuarembó Formation, Paraná Basin, Northern Uruguay). Abstracts 26 Jornadas Argentinas de Paleontología de Vertebrados, Buenos Aires

Quenstedt FA (1889) *Psamochelys keuperina* Jahresh Verein Vaterl. Naturk Würtemberg: 120–130

Rauhut OWM, Martin T, Ortiz-Jaureguizar E, Puerta P (2002) A Jurassic mammal from South America. Nature 416:165–168

Riccardi AC (1991) Jurassic and Cretaceous marine connections between the south-eastern Pacific and Tethys. Palaeogeogr Palaeoclimatol Palaeoecol 87:155–189

Rieppel O (2012) The evolution of turtle shell. In: Brinkman DB, Holroyd PA, Gardner J D (eds) Morphology and evolution of turtles, vol 5, pp 51–62. Proceedings of Gaffney Turtle symposium (2009) in Honor of Eugene Gaffney, Vertebrate Paleobiology and Paleoanthropology. Springer, Dordrecht

Rougier G, de la Fuente MS, Arcucci A (1995) Late Triassic turtles from South America. Science 268:855–858

Santi Malnis P, Colombi C, Kent D, Alcober OA, Martínez RN (2011) Assessing the age of Los Colorados Formation, Ischigualasto-Villa Unión Basin, Argentina. Temporary implication of Coloradian fauna. 4° Congr Lat Paleont Vertebr, San Juan, Argentina. CD-Rom

Smith PL (1983) The Pliensbachian ammonite *Dayiceras dayiceroides* and Early Jurassic paleogeography. Can J Earth Sci 20:86–91

Smith AG, Briden JC (1977) Mesozoic and Cenozoic paleocontinental maps. Cambridge University Press, Cambridge

Sterli J (2008) A new, nearly complete stem turtle from the Jurassic of South America with implications for turtle evolution. Biol Lett 4:286–289

Sterli J (2009) Sistemática y paleobiogeografía de las tortugas continentales del Jurásico de Patagonia, Argentina. Doctoral dissertation, Universidad Nacional de La Plata, La Plata

Sterli J (2010) Phylogenetic relationships among extinct and extant turtles: the position of Pleurodira and the effects of the fossils on rooting crown-group turtle. Contrib Zool 79: 93–106

Sterli J, de la Fuente MS (2010) Anatomy of *Condorchelys antiqua* Sterli, 2008 and the origin of the modern jaw closure mechanism in turtles. J Vertebr Paleontol 30:351–353

Sterli J, Joyce WG (2007) The cranial anatomy of the lower Jurassic turtle *Kayentachelys aprix*. Acta Palaeontol Pol 52:675–694

Sterli J, de la Fuente M (2012) New evidence from the Paleocene of Patagonia (Argentina) on the evolution and paleobiogeography of meiolaniform turtles (Testudinata, new taxon name). J Syst Palaeontol. doi:10.1080/14772019.2012.708674, Cambridge

Sterli J, de la Fuente MS, Rougier GW (2007) Anatomy and relationships of *Palaeochersis talampayensis*, a Late Triassic turtle from Argentina. Palaeontographica Abt A 281:1–61

Stipanicic PN, Rodrigo F, Baulíes OL, Martínez CG (1968) Las formaciones presenonianas en el denominado Macizo Nordpatagónico y regiones adyacentes. Rev Asoc Geol Arg 23:67–98

Sukhanov VB (2006) An archaic turtle, *Heckerochelys romani* gen. et sp. nov., from the Middle Jurassic of Moscow region, Russia. In: Danilov IG, Parham JF (eds) Russ J Herpetol 13 (Suppl.). Fossil Turtle Research 1:112–118

Tasch P, Volkheimer W (1970) Jurassic Conchostracans from Patagonia. Univ Kansas Paleont Contrib 50:1–23

Volkheimer W, Gallego OF, Cabaleri NG, Armella C, Narváez PL, Silva Nieto DG, Páez MA (2009) Stratigraphy, palynology and conchostracans of a Lower Cretaceous sequence at the Cañadón Calcáreo locality, extra-andean central Patagonia: age and paleoenvironmental significance. Cretaceous Res 30:270–282

Wagner A (1853) Beschreibung einer fossilen Schildkröte und etlicher anderer Reptilien Über reste aus den lithographischen Schiefern und dem grünen Sandsteine von Kehlheim. Abh Bayer Akad Wissen Mathematisch-physikalische Klasse 7:239–264

Wood RC, Freiberg MA (1977) Redescription of *Notoemys laticentralis*, the oldest fossil turtle from South America. Acta Geol Lilloana 13:187–204

Zakrszeswski (1888) Eine im Stubensandstein des Keupers gefundene Schildkröte. Jahreshber Verein Vaterl Naturk Wüttemberg Jahrg 44:1–38

Index

M.S. de la Fuente et al., *Origin, Evolution and Biogeographic History of South American Turtles*, Springer Earth System Sciences, DOI 10.1007/978-3-319-00518-8,

P

MIX
Papier aus verantwortungsvollen Quellen
Paper from responsible sources
FSC® C105338

If you have any concerns about our products,
you can contact us on
ProductSafety@springernature.com

In case Publisher is established outside the EU,
the EU authorized representative is:
Springer Nature Customer Service Center GmbH
Europaplatz 3, 69115 Heidelberg, Germany

Printed by Libri Plureos GmbH
in Hamburg, Germany